Algebra Unplugged

by
Kenn Amdahl
and
Jim Loats, PhD.

Algebra Unplugged
by Kenn Amdahl and Jim Loats, PhD.

Copyright © 1995
by
Clearwater Publishing Co
P O Box 778
Broomfield, Colorado,
80038-0778

ISBN 0 - 9627815 - 7 - 6

First printing October, 1995
Second printing June, 1997

Acknowledgments

Special thanks to Cheryl Amdahl and Dee Marcotte for their unrelenting patience. We promise never to utter the word "algebra" in their presence again.

Thanks to Jeff Loats and Scott Amdahl for their kind and careful comments on the first draft. We've recovered from most of them.

Joe Reid deserves a medal for his editing efforts. We've never so enjoyed being slashed by a Red Pen and having all our extra commas, removed.

Larry Shirkey took the photograph for the front cover. Always a pleasure to work with Larry, curator of the free world's emergency supply of bad jokes.

"Art Antics" of Westminster, Colorado supplied the lettering for the cover, quickly and reasonably, after several larger, fancier companies flaked out.

The musicians on the cover are Johanna Parker, Michael Whitfield, Loren Thomas and Kendall Farrington, otherwise known as "Johanna and MLK." They radiate warmth and harmony even when they're not singing.

Access to the alleyway pictured on the cover was made possible by the folks at "Coffee on the Hill" in Denver. Ed Espinoza, Gerald Martin and Gerard Martin could not have been more helpful. Besides great food, and the best coffee in town, these folks have created a kind and gentle community for young urban artists. We also thank the Green Apple Market.

Reed Photo of Denver provided photographic composition for the cover, and treated us like family.

Finally, thanks to all the math teachers and math writers out there who refuse to take our teasing personally. Obviously, we're poking fun at ourselves as well. But if we step over some line, please accept our apology. We only meant to tease those very few overly rigid, stuffy, paranoid, tradition-bound, self-absorbed pedants with no sense of humor. If something we say offends *you*, clearly the fault lies in our own verbal limitations.

Contents

Solving Equations

Graphing

Introduction

Learning algebra is a lot like learning to drive a car, or swim, or ride a bike. You must learn concepts at the same time you are developing skills. There are pedals to push, gears to shift, lights to turn on and off, turn signals to manipulate, and a steering wheel to wrestle. Those are skills.

It's also useful to learn what the brake pedal does, why you need to use your turn signals, what the clutch is good for, and which direction to steer when the car begins to skid. Those are concepts.

You can't learn to drive, or ride a bike, or swim, without practicing. Skills improve with practice.

You can't learn math without practicing either. Most math books acknowledge this as they throw you in the water and encourage you not to drown while you're practicing. Concepts are explained in one terse sentence, often framed in a blue box. If that sentence doesn't make sense to you immediately, you'll have plenty of time to think about it while you're dog-paddling your way through three pages of problems. Sometimes it still won't make sense to you.

It's difficult to learn anything by being placed in the middle of a bunch of technicalities. Some math writers have forgotten this. If they designed a course in bread baking, the students would spend the first three weeks doing nothing but measuring flour. This would be followed by a unit on measuring the temperature of water. Then students would grow yeast. This would take longer than anticipated, so the last chapter, which was to actually bake a loaf of bread, would have to be postponed to the next course.

People who learn best by *reading*, who like to understand concepts before they dive in, have been ignored by the world of math books. English teachers, historians, psychologists and other bright, reading-oriented folks tend to react to traditional exercise-intensive books the way vampires react to Arizona at noon. They're polite, but can usually think of someplace they'd rather be.

This book is different. You won't get much chance to practice. Instead, you'll get an overview of important concepts, such as "don't inhale under water." Our theory is that tidbits like that may prove more valuable if you understand them before you ever get wet.

You'll still need to practice. But perhaps, if we've done our job well, the whole experience will make more sense. It won't be frightening and confusing. Perhaps, while you're taking a traditional class, instead of feeling panic, you'll remember to hold your breath and move your legs like a frog.

Of course, the other students in your algebra class may find that distracting.

What is It?

Algebra is a game. Like many games, it has game pieces, moves, strategies, goals, and its own vocabulary.

Usually, the object of the game is to discover some specific unknown by using available clues. Every time we deduce how many tacos we can buy by remembering how much we've spent, and how much change we probably have in our pocket, we are practicing algebra.

Sometimes the object is to translate recurring events into an equation. Paychecks are figured by multiplying hours-worked by wages-per-hour, then subtracting wages withheld, which are themselves percentages. The whole procedure can be reduced to an equation, which means your boss doesn't have to think very much. Many of us become nervous when our bosses have to think very much.

Unfortunately, the game has been labeled "mathematics." This creates several problems.

First, it is taught by mathematics teachers. These folks have their own way of thinking, and talking, and often have little patience for those of us more comfortable with social studies, carburetors, or literature than fractions. Our confusion baffles them. If they don't love to read, they'll decode whale whistles before they understand someone like me who would rather memorize sonnets all day long than spend 10 minutes identifying the square root of something.

Most of us love games, but aren't crazy about mathematics. We've been bored by math, frightened by it, and made to feel stupid by it. We haven't loved it.

Another problem created by labeling algebra mathematics is this: arithmetic (the math we already know) relates directly to real life. We add apples and oranges, we subtract money, we discover the area of a wall to learn how much paint will cover it. We assume algebra is going to be like that, too.

But algebra is different. While it can, indeed be useful in real life, much of it is simply a game that has no direct relationship to anything we can touch or count. That's fine, in a game. We don't expect checkers to act like apples, or bankers, or astronomical charts. We don't care if $200 is an awfully cheap bribe to get out of jail free, or that it's pretty rare to put 4 houses on a lot before you demolish them all and replace them with hotels. It's a game, for heaven's sake. Sooner or later, successful algebra students figure out that it's a game too. You learn the moves, the strategies and the rules, and accept it for what it is. The elements of the game don't need to relate to anything you can touch. For some reason, it never occurs to anyone to let us beginners in on that secret. We think we must not quite understand yet, and hope the teacher doesn't call on us.

Why Must We Learn This Silly Stuff?

One answer is because it's fun. Your teacher may not mention this, but it's why he chose math as a career. He can't imagine your fear and confusion. To him, it would be like being afraid of ice cream.

There are side benefits to many games. Tennis improves your eye-hand coordination. Racquetball is great aerobic exercise. Golf is meditative and social. Chess teaches concentration.

A side benefit of the algebra game is that it may allow you to become a chemist or electrical engineer. It may allow you to calculate interest rates and design cars and integrated circuits. These are no longer practical applications for us, however, because the people who live in Japan to do those things so well.

Many people say that the thought processes they developed by learning algebra were more useful than any direct application. And, of course, if you want to learn more advanced math, algebra will be a prerequisite.

Beyond that, we are heirs to a long history of science that we can never really understand unless we speak a bit of its language. That language is mathematics, and algebra is its vocabulary. A complete, well-rounded education requires a little algebra, just so you can understand what the biologists and astronomers are saying.

Luckily, it's much easier than you think. The concepts aren't difficult. Besides explaining the concepts, we'll warn you about textbook traditions that have been confusing students for generations. We want to help you relax before your class and we want to tease your teacher out of his stuffiness, and give him permission to put on the big straw hat and play his banjo for you every now and then.

But we don't want to work too hard, and we don't want you to, either. Math is fun. If you notice us getting too serious, feel free to slap this book around until we're back on track.

Someday we'll probably thank you.

Why Do We Think We Hate Algebra?

First, because it uses fractions, and many people never became comfortable with them. Be honest with yourself. You couldn't cross-multiply a fraction if it hopped into your lap and whispered hints into your ear. And neither could I, until Dr. Jim took pity on me.

Second, some of the symbols mean more than one thing. Sometimes an equal sign indicates an equation to be solved, other times it simply shows that two things are the same. A dash might mean subtraction one minute, and a negative number the next. We have to understand what's going on before we can read the symbols, which seems unfair.

Third, there are several unstated rules and traditions no one bothers to explain.

Fourth, they have created a wondrous vocabulary specifically designed to intimidate you. Because you have to understand several of the words before you know what to do with them, the early days don't make much sense, and are therefore boring, which means you don't learn the words, which means you'll be in Big Trouble later, when the teacher assumes everyone knows what she's talking about.

Fifth, they use zero and negative numbers in weird ways.

And sixth, whenever the game becomes messy, with answers that don't make sense or are inconsistent, they add new rules to neaten things up.

The Problem With Math

Teachers want us to believe that math of all varieties is simply a different language to describe reality. They catch us in kindergarten. One apple plus one apple equals two apples. Once we nod our shining little faces and buy that idea, they have us for life.

But they're wrong.

Every apple is unique. No single apple exactly equals any other apple. Certainly no pumpkin equals any other pumpkin, and no zucchini equals any other. No child equals another, no building equals another, no planet, or species, or star, is exactly the same as any other. We always lie when we say one equals one, let alone when we say one plus one equals two. Reality cannot be translated perfectly into numbers.

Mathematics may be a perfect game, but it is never a perfect description of reality.

It is, however, a useful game. One apple may be extremely similar to another. For the purpose of dividing fruit among our friends, mathematics is close enough. For the purpose of figuring trajectories of rockets or the approximate speed of light, it's certainly adequate. In fact, you'll discover a comforting and seductive similarity between reality and mathematics. You may find yourself believing there is a perfect correlation, that reality and mathematics march in lock-step. You might even believe that whatever your formula predicts is true, whether provable or not. Congressmen feel something like this, once they've been in Washington for a while.

It's a dangerous illusion. At one time, excellent mathematicians proved that it was impossible for men to fly in heavier-than-air machines. They have proven that bumble bees can't fly. They established that the world is flat and at the center of the solar system. They have also proved that nothing can exceed the speed of light, that black holes exist, that the universe is expanding, and that lowering taxes will reduce the federal deficit. None of these have been proven by experiment or observation. We believed them simply because the mathematics of the day indicated their truth.

Recently, they've had to create new mathematics that add elements of randomness to explain the motion of water in a waterfall and the pattern of maple trees. They've had to admit that the math we've been using (and that they still teach) simply can't describe those situations perfectly.

But, of course, sometimes we need to simply count things. It doesn't matter if one apple doesn't exactly equal another. Each of the trick or treaters needs something in his sack, we have a limited supply of chocolates, and we need to know how to distribute them.

Step one in learning algebra is to understand that each element in it will not relate, exactly, to anything in the known universe. Your teacher, and your Real Algebra Book, will use examples from life to explain each concept. But these examples can only give you a feeling for how the game is played or what a strategy might be. If you cling to them too tightly, you'll panic when you realize that the example wasn't perfect.

Like any game, algebra must be internally consistent. The game is played within specific limits. Moving checkers off the board and onto the carpet is forbidden. Football players can't throw pumpkins or use helicopters. In algebra, you can't divide by zero. You can protest all you want and say it doesn't make sense. But the game doesn't work as well without that rule and others like it. It's less consistent and predictable, and therefore less useful. It has nothing to do with counting apples, or holes in the ground, or your checkbook balance. It doesn't have to make sense in the real world. You can't divide by zero while you're playing algebra.

Chess

Dr. Jim stared into the distance thoughtfully. Something was bothering him. We were just finishing breakfast, discussing the algebra book we intended to write together. It looked like a big project to me, because I didn't know anything about math. It looked like a big project to Jim, too, because writing intimidated him. But for the moment, we were talking about the game of chess.

"I never became good at chess," he said. "Oh, I learned how the pieces move, and the rules, but I always wondered about something."

I nodded and drank my coffee. Chess had been an important part of my life during high school. Perhaps if any of the girls had been willing to talk to me, I might have had less experience with knights and bishops. Jim fidgeted, trying to think of a way to phrase his question. He almost seemed embarrassed.

"I know that chess is an old game," he said. "And the pieces represent various characters in ancient civilizations. I mean, you've got the queen and king, and knights and bishops and pawns and castles. So, if you really understand the game, when you move your knight, are you thinking, 'How would a real knight move in this situation?' Does it help to keep a bishop's vows in mind when you move one?"

I almost laughed out loud at this absurd idea, but caught myself when I realized he was serious. He felt there was some deeper mystical aspect to chess that he'd never known how to learn. He sensed a magical relationship between medieval queens and the game piece with the crown, yet couldn't figure out how to make the wooden game piece move in a queenly fashion. Where was her majesty, her regal bearing?

"No, Jim," I said slowly. "I don't think anyone considers the game pieces to be anything more than symbols for the various possible types of moves. Bishops move diagonally. If we need a piece that moves diagonally, we might consider moving our bishop. It's possible that as we move it across the board we might imagine flowing robes and a priestly demeanor. But probably not. We get so wrapped up in the game itself that any relationship to the real world fades away. Sorry."

"Not at all!" he said, his face relaxing into a smile. "What a relief! All the time I was trying to learn chess no one would explain how a knight is like a knight or what makes a castle similar to a real castle. This is great! You're telling me that chess is just like mathematics!"

I shook my head. "No, Jim, I don't think you get it. In chess, the pieces don't represent any real thing. It's just a game. In math, the numbers and symbols do represent real things. Apples, or oranges, or a rocket's path. We use math to design things, to predict things, to explain reality. See, it's different."

He stared at me. "You don't really believe that, do you?"

"Well, of course I do. That's why people have to learn algebra and the other forms of math. So they can operate in the world of science or engineering."

"Yes, yes," he said, waving his hand as if shooing mosquitoes. "It's useful in the real world. But that doesn't mean that the numbers or symbols or rules relate to anything in reality. Surely they explained that to you in elementary school."

I was quite sure no one had ever explained this to me. "I don't exactly recall..."

"Mathematics is like a little engine," he said, "with lots of gears and spinning parts. The engine is inside a black box. All the parts work together. They fit perfectly. This part turns that part, which pushes a rod into a slot, which winds a spring which turns the wheels beneath the box.

"We feed information into the black box, and the little engine starts whirring and sputtering. The information may come from the real world, but each gear and pulley inside that box doesn't need to represent some real thing.

"You can put the engine in your car and drive to the movies with your girlfriend. The engine doesn't understand the movie, or the girlfriend, or even the street beneath its wheels. It doesn't need to. All that is important is that each part mesh with the others within the box, that they work together to transform your information into motion. It's exactly like chess."

Great Lies About Mathematics, by Dr. Jim

You probably bought a bunch of "lies" about math or you wouldn't be reading this book. If you continue to believe these lies, it will be harder to learn math. Here are some of them:

- *There is only one way to get the answer.*

    When you're stuck, you'll be even more frustrated thinking that you have to find THE one way to work the problem.
- *All problems can be solved by using a step-by-step method.*

    Steps are important, but overrated. That's why we wrote this book about concepts and not step-by-step methods.
- *There is a math gene some people have and others don't.*

    Guaranteed to make you feel helpless often.
- *Math is hard, too hard for most people to learn.*

    Test data from across the world tells us this isn't true. But many American mathematics *classes* are organized in ways that don't make sense to students. So our culture has believed this lie for many years.
- *If a math problem takes more than 5 or 10 minutes, it is impossible.*

    Any really interesting math problem will take much longer. Typically, math books contain exercises, which are dull and dulling to those who work on them. Real problems can take far longer and are more interesting.
- *Math is mostly memorizing.*

    This is one of the most deadly lies, because people

believe it and get through the first couple of years of algebra using it. When they arrive in calculus class or any other occasion where being able to think is required, they discover that they don't have the understanding that is needed.

- *Only geniuses are capable of creating or understanding formulas and equations.*

If this is true, then you'll be a genius after reading this book.

These "lies" get in your way. On the other hand, too much "truth" is just as bad.

Imperfect explanations give mathematicians the creeps, especially when they see them in writing. But perfectly accurate, complete, irrefutable explanations tend to be boring.

Creative analogies (comparisons to familiar things) are rare in print and common in spoken math. Written math is formal and rigid, while spoken math is usually less precise. Only skilled math types can listen to spoken math and discern errors and omissions on the fly. In my teaching, I try to use lots of analogies and some sloppiness so I can draw my students in without the rigidness of written mathematics.

The lack of analogies is a tradition in math books. Any analogy is, of course, not the real thing. It will have some aspect that is not accurate (truthful). But the fine points aren't apparent until a student has developed some sophistication. That process can be greatly speeded up by using analogies. For example, in explaining how negatives work, there just doesn't seem to be ONE good model that we can use for adding and multiplying. Each analogy "lies a little" or at least isn't always useful or extendible. So, math books don't use them at all.

This stiffness and lack of analogies makes people

nervous. They feel self-conscious and inadequate. They avoid math.

As you overcome your superstitions, you'll realize you can relax. You can play with math. You can enjoy it. You can learn it.

========

The Game in General

Algebra is a tool kit for solving mysteries. Using whatever clues we find, we translate the mystery into numbers and symbols. Then, by employing specific strategies, we extract information from the numbers and symbols. After we've manipulated the numbers in accordance with the rules, we translate the result back into real-world terms. That's algebra.

Notice the three discrete steps:

1. Translating the real-world situation into numbers and symbols is one step.

2. Manipulating these is a second, completely different activity. This activity is the game of algebra.

3. In the final step, we translate the solution back into the real world and see if it makes sense.

There are two objects of this game. Much of the time, the goal is to pull the mask off an "unknown." Other times, the goal is to translate a repeating pattern into math. Algebra is the language of patterns.

Practical problems begin with a situation that can be described in words and sentences. Often your Real Algebra Book will skip this part altogether, in the interest of saving space, and proceed directly to the manipulation of the numbers and symbols. In the real world, algebra problems won't appear ready-made. They'll be disguised as mysteries, confusions, complications and dilemmas.

It will be up to you to translate them.

First we must identify the mystery, which we call "the unknown."

"Subtract some number from 20 and you'll get 15."

"Some number" is the unknown. Traditionally, letters near the end of the alphabet indicate unknowns. The most common letter to use is "*x*." In our example, we replace "some number" with "*x*."

"Subtract x from 20 and you'll get 15."

We have identified the unknown.

At this point we employ a subtle shift away from our normal thought process. Usually, results come at the end of a series of logical steps. If I plant the seed, water the sprout, and pull the weeds, then I'll be able to harvest my watermelons at the end of the summer. If I follow the map, I'll reach my destination. Arithmetic behaves like that. We know all the numbers we're going to add up. The unknown is the answer at the end.

In algebra, the unknown often springs up in the middle of events: "If I plant the seed, water the sprout, *and do something else that I've forgotten*, I'll be able to harvest my watermelons at the end of the summer." Or: "If I follow the map, *including the part that's been torn off*, I'll reach my destination."

If we fill in that missing part, it's a true statement.

The true statements of algebra are "equations." An equation is a math problem with stuff on both sides of an equal sign. It doesn't matter that some of the items on each side might be unknowns. We arrange the game pieces into a formula that will be true if we substitute the correct answer for *x*:

$$20 - x = 15$$

This step is often eliminated from algebra books, probably because it makes it possible for you to use algebra in real life.

Next, we play with the equation, being careful not to alter its truth. We twist and pull until all its numbers sit on one side, and its unknown sits alone on the other side.

Then we complete the arithmetic on the number side to learn the unknown. The unknown has become known. It's our answer.

In step three, we use the answer in the real world to build a bridge, or cut up some pizza, or sew a dress. Your Real Algebra Book will skip this part as well. They rarely let you build the bridge or eat the pizza. Too bad. That's the reason you're playing the game in the first place.

Solving Equations

Years ago people determined how much something weighed by putting it on one end of a scale. In those days, a scale was like a teeter-totter, a long board balanced in the center. They kept various weights handy, each one labeled. If a one-pound weight balanced your sack of peanuts exactly, Mr. Grocer charged you for one pound of peanuts.

An equation is like a balanced scale. The stuff on the left side of the equal sign weighs exactly the same as the stuff on the right side. To solve an equation, we have to figure out what's inside any unmarked boxes.

Whatever tricks we try, we are guided by one rule: we can't let the scale tip away from perfect balance. That's the game. Keep the scale dead flat. The left side must equal the right side at all times. You can do anything you want, if you obey this rule. Tap dance on the scale, drive your car over it, send the whole works down Niagara Falls. Just don't ever let the scale tip away from equality.

Our first instinct is to suspect there's not much we can do to an equation that won't send it tumbling. But there are two things we can do to either a scale or an equation: We can replace one item with an equivalent item. And we can do the same thing to both sides without tipping the balance.

Replace numbers with their equivalent

We can describe any number in many ways. The number "16" for example can be described as 4 times 4, or 16/1, or 32/2, or 64/4. It can be described as (15 +1), (40 - 24) or 16,000,000 ÷ 1,000,000. One strategy for solving equations is to replace a number with another version of the same number. If we have a one pound weight, and a one pound sack of peanuts, it doesn't matter which one we use on the scale. So, if we see

$$(15+1) + 2/3 = x$$

we can replace (15+1) with 16 and the scale won't even quiver. They're simply two different ways to say the same thing. Unfortunately, some substitutions don't do us much good, either. We could replace 2/3 with 4/6 and we will not have affected the truth of the statement. But does it help us? Maybe, but maybe not. Obviously, it's going to take a little practice to decide which of the many variations of any number might do us some good. This is where an algebra class will help you. Any good teacher will give you many opportunities to flounder around helplessly in search of experience. Just remember this: you can always replace a number with something that is exactly equal to it.

Do the same thing to both sides

If the scale is balanced to begin with, and you add a penny to each side, it will still be balanced. If you subtract a one-pound sack of peanuts from each side, it will still be balanced. In fact, you can do anything you want to one side of an equation as long as you do the same thing to the other side. Multiply, divide, add, subtract; it doesn't matter. Double the weight on both sides of the scale and it will still be balanced. That is, multiply both sides of the equation by 2 and you will not have impeached the truth of the statement. Divide both sides by 2 and you'll be back where you started, and still balanced.

You won't affect the balance if you do the same thing to both sides of an equation.

The Goal

There are two common ways to win at algebra. One is to identify the unknown. When we can say "x equals 43 compassionate Republican senators," we have solved the equation, although in this case we might want to check our arithmetic.

The other is to identify an equation that represents a pattern. Perhaps we've been given a list of some sort, in which one column relates to the others in some common way. When we can say "The number of new social programs each year equals the number of Democratic representatives squared," we have reduced the list to a formula and saved a lot of paper. A formula expresses this more efficiently than a chart with several hundred possibilities.

You need to know which result you're aiming at.

If you're trying to solve an equation, you hope to arrive at a simple answer. But sometimes, it won't be possible. In those cases you win when you recognize the nature of the problem.

Some equations are true for all numbers. Doesn't matter what you plug in as the unknown, it'll always work.

$$x - x = 0$$

You can replace x with any positive or negative number, any fraction, any dairy product, any small appliance, and most small children. It will always be correct. Not very useful, but correct.

Some equations are never true, no matter what number you try:

$$x + 1 - x = 0$$

You are done with this problem when you recognize that it can't be solved.

And some unknowns resist a simple definition. The closest you may get is a simpler equation. This is not a failure on your part. Sometimes you get a touchdown, sometimes you settle for a field goal.

Our optimistic assumption is always that, by being persistent, we will discover the identity of the unknown. With that cheerful attitude, we begin manipulating our equations, using the tricks and strategies we've learned. But, until we focus on our goal, we're simply conducting arithmetic exercises, madly squaring numbers and multiplying fractions with no clue as to why anyone would want to spend a pleasant Saturday afternoon this way.

The big strategy is obvious, if you think about it for a moment. You want to get x isolated on one side of the equation. You want to be able to say "*x* equals . . . whatever's on the other side of the equation."

To solve equations, we do identical things to each side of the scale, and substitute equivalent numbers. That's the whole game. It seems more complicated than that because it's not always easy to translate a problem from "life" into math. It requires many words and concepts to describe all the possibilities. To survive an algebra course, you're going to have to learn the language.

Often it helps to hear two different explanations for a word or concept. We're here to provide that second explanation. Ours may not be an improvement on your Real Algebra Book's explanation, or on your teacher's. But it will probably be different.

There is no perfect "order" to learn these words and concepts. Like any language, it's hard to explain one word without using other words that may also be new. Some won't make much sense to you until you use them. Don't worry too much about the order of concepts. We're just going to whimsically wander through the landscape, tossing out an occasional pretty word, in whatever order they occur to us. There is no "right" way to do this. There is only "the way it's always been done," and, now, our way.

The Game Pieces

Many games come with a fixed number of pieces. We use 52 cards to play most card games, 32 chess men for chess, and 4 horseshoes make a set. Two guitars, a bass, a drum, and a lot of depression are required to play the "alternative rock band" game.

Depending on our situation, we play algebra with different sets of numbers.

There are many kinds of numbers. For example there are fractions, decimal numbers, and whole numbers. Some numbers fit into several of the categories below. This list is arranged in a specific order. Usually one learns about the numbers in the order below. But more importantly, each of these sets of numbers includes all the numbers in the sets listed above them.

Sesame Street numbers: Prior to 1985, these were the numbers 1, 2, 3, 4, 5, 6, 7, 8, 9, 10, 11, and 12. You remember "The Count." By 1985, the Count had learned how to go all the way up to 20. (Note: Mathematicians do not use this classification.)

Counting numbers: As you might guess, these are the numbers you count with. 1, 2, 3, 4, 5, 6, 7, 8, 9, ... (Those little dots mean that the list is too long to type out.) Some books will refer to these as the **natural numbers**.

Whole numbers: These are the same as the counting numbers except that zero is also included. So this list goes like this: 0, 1, 2, 3, 4, ... The word "whole" is used to exclude the fractional numbers. This is the kind of numbers you would use to describe how many complete banana cream pies you can to eat, or how many roller coaster rides you enjoyed last night. Some books will refer to *these* as the **natural numbers**. There is some disagreement among text book writers as to whether zero is a natural number or not.

Integers: The integers include all the whole numbers and all of their negatives too. Here's the list:
$$... , -3, -2, -1, 0, 1, 2, 3, ...$$
Another way to list them is to use the plus or minus sign,(±):
$$0, \pm1, \pm2, \pm3, \pm4, \pm5, ...$$

Note, for example, the number 5 is a counting number, a whole number, and an integer.

Rational Numbers: These are the fractions. Some examples of rational numbers are:

$$\frac{2}{3} \qquad \frac{34}{9} \qquad \frac{-4}{3} \quad \text{and} \quad \frac{21}{-4}$$

The official definition say these are exactly the numbers that can be expressed as the ratio of two integers where the bottom number (the denominator) can't be zero. The word rational refers to"ratio." This is a complicated bunch of numbers. Even deciding when two of them are equal involves some work. A **mixed number** contains both a whole number and a fraction, like:

$$4\frac{2}{3}$$

Because this number is equal to

$$\frac{14}{3}$$

it is a rational number. When rational numbers are represented in decimal form, the portion of the number to the right of the decimal point either stops, like:

$$\frac{3}{4} = 0.75$$

or it repeats in blocks, like:

$$\frac{35}{11} = 3.181818...$$

Real numbers: A real number is any number that can be represented as a decimal number, positive or negative, repeating or non-repeating. The real numbers can be used to label every point of a line. In fact, when you draw a number line, it is customary to think of each point as representing a number and vise versa. All the numbers we have discussed so far are included in the real numbers. Some real numbers are not rational. They are called irrational numbers and one example is π. The "square root" of two is another. They resemble rational numbers, except that when you express them in decimal form, those digits go on and on and on, never repeating, and never falling into a repeating block of digits.

Complex numbers: Complex numbers are a combination of real and imaginary numbers. That may take a bit of explaining.

Rather than stop when things actually made sense, mathematicians got carried away and created an even bigger set of numbers.

You can't multiply anything times itself ("square it") and get a result that's a negative number. So, the negative numbers don't have "square roots." It would be handy if this were not the case. So, they invented a kind of number that would satisfy this requirement. Appropriately, they call them "imaginary numbers," represented by a small "*i*."

A math teacher might explain that a complex number is a number of the form $a + bi$, where a and b are real numbers and i can be thought of as the square root of -1. In other words, complex numbers have a real component and an imaginary component.

The good news is that the craziness stops with the complex number system. If you study college algebra, you may encounter these numbers. They are rarely used outside of scientific settings.

If you like different kinds of numbers, a fascinating literature awaits you. You can discover numbers that are perfect, amicable, palindromic, figurate (like square and triangular, etc.), abundant, etc. You will be following in the footsteps of folks who began looking at these things as early as the Egyptian and Babylonian civilizations. Enjoy yourself.

The Moves

Checkers can move forward diagonally to the left or right, or can jump an opponent. In poker, you raise, call, fold, draw, etc. In football, you run into people. Every game has "moves" of some sort.

There are four basic "moves" in algebra: addition, subtraction, multiplication, and division. These "moves" occur after the words have been translated into numbers and symbols during the manipulation of equations. The "rules" of the game tell us what order these must be done in.

The "moves" are called "operations." An operation is an activity, like adding, or multiplying. The "order of operations" is the rule that instructs you which to do next in a problem.

This may be a welcome surprise to you. There are no new multiplication tables to memorize or anything as difficult as long division. Algebra is a way to use your old skills. You aren't starting over. You're moving on.

The Strategies

There are ten basic strategies in algebra:

1) Your overall strategy is to manipulate the equation, without affecting its truth, until you have a single unknown on one side of the equation and an arithmetic problem on the other. Once you complete the arithmetic, you'll have the answer.

2) You can do the same thing to both sides of an

equation without affecting its truth. If you add 30 to the left side, and 30 to the right side, you have not affected the balance.

3) Reversibility. If you double a number, you can find your original number by halving the new number. If you add 10, you can reverse that by subtracting 10. If you square something, you can reverse it by taking its square root. This is a lot handier than it sounds.

4) It doesn't matter what order you add things, if that's all you're doing to them. So, sometimes you can change the order of adding things to gain a strategic advantage.

5) It doesn't matter what order you multiply things, if that's all you're doing to them. It may be useful to change the order in which things are multiplied.

6) When you need to multiply a number times the sum of several other numbers, it doesn't matter if you add them up first, then multiply, or if you multiply each of the individuals first, then add the results.

7) Rule #6 also applies to division. If you are dividing the sum of a bunch of numbers by one number, you can complete either the addition first, or the division.

8) Multiplying anything times "one" doesn't change it. But "one" can be written many different ways, and the results of the various ways will each look different, although they're not.

9) Dividing anything by "one" doesn't change it.

10) If the answer to a multiplication problem is zero, one of the numbers in your multiplication problem must be zero. This is also a lot handier than it sounds. Sometimes, rearranging an equation into a multiplication problem which equals zero simplifies your life. This is called "factoring."

That's it, in a nutshell. These are the strategies you'll use to reshape equations to your specifications. Notice that four and five are nearly identical, as are six and seven. So, there's really only about eight techniques. It hardly seems complicated enough to devote a whole course to, does it?

The Rules

The primary rule of the game is that you must do things in the proper order. This is called the "order of operations" and involves a few simple principles that you use over and over. Don't worry that you may not understand some of these words yet. Unless otherwise instructed, you do things in this order:

1. You deal with things in parentheses first;
2. Then you deal with exponents;
3. Then multiplication and division problems;
4. Then addition and subtraction problems.

It may help you to remember a simple sentence that uses the first letter of each operation: **P**lease **E**xcuse **M**y **D**ear **A**unt **S**ally. That is, **P**arentheses, **E**xponents, **M**ultiplication, **D**ivision, **A**ddition, **S**ubtraction. Of course, you can make up your own little sentence, like, "**P**ale **E**lvis **M**ay **D**ance **A**t **S**avannah," if that would be more memorable. Or "**P**olite **E**lves **M**ake **D**andy **A**fternoon **S**andwiches."

Notice that multiplication and division are part of the same step, and addition and subtraction are part of the same step. Complete multiplication and division problems from left to right as you encounter them. Do the same with addition and subtraction problems. Don't do all your multiplication problems first, then your division problems, even though the Aunt Sally order might suggest you should.

Given a choice, work from inside out. If one parenthesis is nested inside another parenthesis, work on the inner one first. Within parenthesis or other groups, follow Aunt Sally's order.

Doing things in the correct order is the primary rule of algebra. A sentence makes little sense if you try to read it backward. Backward it read to try you if sense little makes sentence a.

A couple of things you might try are simply forbidden. This is necessary to make sure the game remains consistent. Unfortunately, you need to understand quite a bit of math before you recognize their danger. So you may want to protest some of them as being silly when you encounter them. There ought to be a way to divide by zero, for example. What harm can it possibly do anyone?

It's a sensible question. It's not immoral, difficult, or fattening. The short answer is this: It just messes up the game, so we all agree not to do it.

A more subtle danger occurs when something usually works. You learned to cancel fractions in fourth grade, for example. In algebra, you'll still get to cancel fractions but it won't always work. If you believe math operates according to cosmic laws, you'll feel betrayed: They've been lying to you for years.

But they haven't been. They've been playing a simpler version of the sport. The rules of beach volleyball change depending on who brought the ball. It doesn't bother us because we know we're only playing a game.

Try to think of algebra as beach volleyball with numbers.

The Multiplication Sign

Most of us learned that multiplication should be represented by a little "x." Ten times ten always looked like this:

$$10 \, x \, 10$$

This becomes a problem if we're dealing with problems that include a lot of unknown numbers represented by "x." Our multiplication sign and our unknowns look nearly identical.

To prevent confusion, we replace the multiplication sign with a dot. Now our problem looks like this:

$$10 \cdot 10$$

The dot isn't some fancy new procedure. It just replaces the multiplication sign.

To make things even neater, whenever it doesn't cause confusion, any two distinct items should be multiplied together if they adjoin.

$2x$ means "2 times x."

$23(4+5)$ means "23 times the sum of 4 plus 5."

Notice that 23 still means "Twenty-three." It does not mean "2 times 3." That would be absurd, wouldn't it?

$46xyz$ means "Forty-six times x times y times z."

Zero

Zero is easy to understand when you're counting apples. If you have 10 apples, and you eat 10, you'll wind up with a stomach ache. And zero apples.

One plus zero equals one. If you have one apple, and no one gives you another, you're stuck with one lousy apple.

One minus zero is also easy.

Zero doesn't do much when you're adding or subtracting. But it becomes all-powerful in multiplication and division problems.

Zero times anything is zero. Zero, like Zorro, is a powerful character, able to vaporize any number or unknown simply by multiplying by it. A million times zero equals zero.

It also overpowers division problems. Zero divided by anything is still zero. It doesn't matter how many checks you write, if there was zero money in the account to begin with, all your creditors will be equally unhappy.

Worse still is trying to divide another number by zero. We can divide our pizza into four parts, and no one will be confused. We can even imagine dividing our pizza into one part and eating the whole thing ourselves. But we can't imagine dividing it into zero parts.

So we don't. We simply declare dividing by zero to be illegal in our game, and we move on. You can't divide by zero.

Repeat after me: "You can't divide by zero."

One

One is such a nice, quiet number, it often gets overlooked. But it may be the handiest number in algebra.

In addition and subtraction, it affects things by a small amount. But in multiplication and division, it does something even more remarkable: It does nothing.

Any number times one equals the original number. You can multiply something times one all day long and never change it at all. What makes this valuable is that you can write the number "one" so many ways. By employing different varieties of "one" you can convert numbers to various fractions, or transform existing fractions, without changing the size of any of your original numbers. "One" is a remarkable translation tool. It works like this:

Any number divided by itself equals 1. So any fraction with the same number as numerator (the top number) and denominator (the bottom number) equals 1. When you multiply it times anything else you don't affect anything except the way it looks. But sometimes, that change is the key to solving a problem.

$$\frac{435}{435} = 1$$

Numbers are also unchanged when you divide them by 1. A million divided by 1 is still a million.

This is handy because any number can be transformed into an equivalent fraction simply by making "1" the denominator:

$$12 = \frac{12}{1}$$

You can use this fraction anywhere you have a twelve and not change anything. So, if you see a strategic advantage to transforming twelve into some fraction, it is frighteningly simple to do. Divide it by 1.

Negative Numbers

Nothing in reality is exactly like negative numbers. That's the hardest concept. Negative numbers exist as game pieces in algebra. We can understand them, in an abstract sort of way, but every time we try to visualize them, we fail. Worse, we believe we understand them before we do, and move on too soon. We see

$$20 - 30 = (-10)$$

and say, sure, no problem. That's similar to many things.

A negative balance in your checking account? Yes, that's good. You must add a hundred dollars of real cash to offset that hundred dollar overdraft the bank keeps reminding you about. But, if it weren't such an abstract concept, you wouldn't have written all those bad checks to begin with.

A hole in the ground? Yes, we need to add so many shovelsful of dirt to fill it in. But who cares? Few of us spend a high percentage of our days calculating how much dirt we need to put back into a hole to bring it level.

You smile, you nod patronizingly. You don't believe it's a tough concept. Combine negative 8 and regular 8 and you'll get zero. Ten combined with negative 8 equals 2. No problem.

This arrogance will get you in trouble. At some point, someone will propose a problem something like this:

You have 3 apples. How many apples must you subtract to end up with 8 apples?

Those of us who cling to the belief that algebra relates to real life will howl. Can't be done, we'll say.

But, in algebra, if you subtract a negative 5 apples from 3 apples, you get 8 apples. Subtracting a negative 5 is the same as adding a positive 5. That is,
$$3 + 5 = 3 - (-5)$$
These negative numbers can be used everywhere a regular number can be used, but you get goofy answers. The confusion is compounded by the fact that a dash indicates a negative number, but it also indicates subtraction. So you have to make sure you know what that little dash means in any problem.

Picture negative numbers as weights in a hot air balloon. The more of them you add to the basket, the lower the craft will drift. If you subtract these weights, the balloon will ascend.

Positive numbers might be balloons of helium. The more of these you add to your craft, the higher it will fly. Subtract them and you'll sink.

If your hot air balloon is hovering at 3 feet off the ground, what must you subtract to reach 8 feet off the ground? You must subtract weights. You must subtract negative numbers.

Alternatively, you could add helium balloons. That is, you could add positive numbers.

Negative numbers are trickier than the simple two lines they will receive in your Real Algebra Book. They aren't really like weights in a balloon, or dirt in a hole. They are special creatures that exist only in this fairyland of math. No simple analogy will be consistent in every case. They behave exactly the opposite of positive numbers. When you add them to something, you wind up with a smaller number. When you subtract them from something, you wind up with a bigger number. When you add two of them to each other, you get a number smaller than either one of them. When you multiply two of them together, you get a positive number. When you multiply a negative number by a positive number, you get a negative number.

Does that make sense? Of course not! Is it reasonable for you to understand this after reading it once? Absolutely not. But is it difficult? No. It's just unusual, if you're not used to thinking in these terms. Every single person who has become comfortable with negative numbers has wrestled with them and subdued them in their own way. You'll have to do that too. If you simply memorize the preceding paragraph, you'll improve your grade. But you'll still have to fool with negative numbers for a while to become comfortable using them.

We'll talk more about negative numbers, your teacher will talk about them, and so will your Real Algebra Book. The important thing to remember for right now is that negative numbers are not trivial, and that understanding them in a vague sort of way isn't good enough. You have to know how they behave or you can't play the game. Confusion about negative numbers prevents many people from succeeding at algebra.

But you won't be one of them.

Losing Weight

Braindead, the algebra student, recently joined a club for the metabolically impaired and the willpower deficient. He was a little nervous about attending his first meeting. He wore a triple-extra-large sweat shirt that still felt snug around his ample waist. The room was full of corpulent folks with wealthy belt sizes.

"Yo, Porky!" they yelled as they saw him. Braindead smiled, happy to be among friends.

The only thin person in the room was Miss Pounder, an energetic woman with very short, nearly white hair. She stood at the front of the room, and everyone got quiet.

"Welcome to MIWD!" she shouted, pronouncing the letters as "Mee-wod."

"Glad to be here!" the crowd shouted back in unison.

"What time is it?" she shouted.

"Badge time!" the crowd responded gleefully.

"That's right! Line up by the scales and let's see how we did this week!"

Braindead stood in line. One at a time, members stood on the scale. When someone lost 5 pounds, they received a silver medal, about the size of a half-dollar.

"It doesn't matter where you start," Miss Pounder explained for the new members. "It's how many badges you earn!"

Some folks had chests covered with dangling silver medals.

"No! There's got to be some mistake!" a woman screamed. "I haven't eaten a thing all week!"

"Now, Mrs. Jeffries, we've been through this a

34

hundred times!" Miss Pounder was kind, but firm. "The scales don't lie! Now give me back one of your badges!"

"But I swear! How can I have gained 5 pounds if I haven't eaten?"

"Maybe you've been sleepwalking into the kitchen again, Mrs. Jeffries! Or maybe Mr. Jeffries has been injecting you with gravy while you were distracted with one of your favorite game shows! Didn't that happen to you just last summer?"

"But, but..."

"Give me the badge!"

Mrs. Jeffries' hand shook as she took off the badge and dropped it into Miss Pounder's hand. The hand closed like a bear trap around it.

"I'll just keep it for you," Miss Pounder said. "You can earn it back next week!"

Mrs. Jeffries stood beside Braindead.

"I don't get it," Braindead said. "The badges. They're so pretty. But... it's very confusing to me."

Mrs. Jeffries nodded. "It's a hard concept for some of us. For every 5 pounds you lose, you get a badge."

"But you didn't lose any weight."

"That's right, smart boy. I gained a little. Aliens. That's what I think. I think aliens kidnap me and feed me while I'm on their ship. In fact, I'm sure of it! Miss Pounder! Miss Pounder!" She waved her hand and yelled excitedly, but Miss Pounder did not look up. "Oh well," she said, "I guess I can explain it to her later."

"So when you gain weight, they take away one of your badges?"

"Yeah."

"But I don't have any badges. What happens if I gain 5 pounds?"

"Scarlet badge," Mrs. Jeffries said. "Every 5 pounds above your base weight buys you a scarlet badge."

Braindead glanced around the room.

"So those people with many silver badges have lost a lot of weight?"

"Bingo."

"And that guy over there?"

"The guy who looks like his coat was made from scarlet badges? Sad story. Came in to write an article on the program. Wasn't even heavy. But, after a few weeks, he was hooked. All this talk about food and losing weight just seemed to give him an appetite. Porked right up. Another month or two and he'll need a crane to lift him off the couch."

"I'm sure glad I joined. This really sounds like my kind of place," Braindead's face was aglow. "Do they really have cranes?"

How Negatives Work in Multiplication

When you multiply a negative times a positive, your answer will be a negative.

When you multiply two negatives together, your answer will be a positive number.

After she gets home, Mrs. Pounder makes notes about her night at MIWD and the total amount of good that she was able to do. She has to make her weekly report to headquarters.

"Well, for six people I gave out three silver badges each.. If each silver badge means someone lost five pounds, then three silver badges is like a negative 15. I'll have to multiply that by the number of people who received three badges to determine how much weight they lost altogether." So she writes this:

$$6(-15) = -90$$

"For three people I gave out three red badges each." A red badge means weight gain, so she represents them as a positive number. She writes this down in her book:

$$3(+15) = 45$$

"For five people I took back two silvers." Since taking back a badge is the opposite of giving someone a badge, she decides to consider those people as negative numbers. She writes:

$$-5(-10) = 50$$

"For two people I took back three reds." She writes:

$$-2(+15) = -30$$

It was a very good day.

The "Minus Sign"

The little dash we once knew as a "minus sign" means three different things. The three meanings are similar enough that they slide past us without much contemplation. Later, our muddy thinking catches up to us. Hopelessly lost and frustrated, we may quit the class in despair, and give up on becoming a doctor or engineer. This is why there are so many attorneys in America. No math.

When we are young, the little dash means subtraction. Ten minus 3 equals 7. This concept, by itself, doesn't bother us.

$$10 - 3 = 7$$

The dash can also mean a direction, rather than an action (like subtraction). A lot of math consists of making little pictures, like maps, to represent something. In these pictures a plus sign means move one direction. A dash means move the opposite direction. Once again, no problem. If "3" means move three places to the right, then "-3" means move three places to the left. And -(-3) means move three places to the right. If plus means up, minus means down. The end result still seems a lot like simple subtraction.

The dash also designates a negative number. If you have zero money in the bank, and you write a check for 80 dollars, your account now contains a negative 80 dollars:

$$0 - 80 = -80$$

This is so seductive! We used subtraction, and our result is a negative number.

Don't be fooled. Subtraction is an activity, like swimming, multiplying, or dancing. A negative number, on the other hand, is a thing, not an activity. It's one of the game pieces, a character in the play. We might have reached that negative 80 in many other ways.

For example, if last week you wrote a bad check for a hundred dollars, then you're probably not ready for this book. Even so, you wound up with a balance of negative 100. Perhaps today you deposited 20 bucks. Now you have that same balance of negative 80. But you got there by adding to your balance.

You can do almost anything with negative numbers that you can with positive numbers. You can add them, subtract them, multiply them, divide them, or square them.

But you need to beware of that little dash. It can be telling you to subtract, or giving directions on a graph, or identifying a negative number. It's not a simple "minus sign" anymore. Some math teachers won't realize that this is confusing to you.

It will be especially confusing when you face negative numbers.

Adding negative numbers is a lot like subtracting. Ten plus a negative 3 equals 7.

$$10 + (-3) = 7$$

You might think of negative numbers for a moment as little silver badges given to the metabolically impaired to represent weight loss. One badge equals 5 pounds lost. So, if you started off at 300 pounds, and you're wearing 3 badges, we might deduce you now weigh about 285. The more badges you add, the less you actually weigh. Negative numbers work like that.

Subtracting negative numbers seems like an odd thing to do. It's a lot like taking back badges already won. If you weighed 300 pounds last week, and were wearing 6 badges, we know you started out weighing 330. But this was Super Bowl week, there were nachos, there were milk shakes, perhaps there were aliens. This week they take away 2 of your badges. We have subtracted a negative 10 from your last weight of 300. Now you weigh 310.

$$300 - (-10) = 310$$

Multiplying negative numbers with positive numbers is like trying to figure the weight loss of several dieters. Perhaps 6 dieters are each wearing 3 badges. Each of them has lost 15 pounds. What was the total change in their weight?

$$6(-15) = ?$$

Altogether they lost 90 pounds. 6 times negative 15 is a negative 90.

$$6(-15) = -90$$

You might guess how multiplying two negatives works. If multiplying a negative times a positive gives you a negative, and you know that negative numbers act contrary to positive numbers, then multiplying two negatives will give you a positive. Negative 2 times negative 3 results in positive 6.

Negative numbers are just that quirky.

We can divide up our silver badges among dieters. If we have 6 badges to distribute to 3 dieters, they might each receive a couple, and each would represent a negative 5. A negative divided by a positive yields a negative.

Going back to our concept of the contrariness of negative numbers, it makes sense that if you divide a negative number by another negative number, you'll probably get a positive number as your result, even if you can't stretch your story about badges to make a tidy little example from the real world.

As it turns out, for the game to be consistent, this is what must happen in each of these situations. It doesn't necessarily make sense, and we may not be able to come up with a cute analogy that will fit every case.

But, of course we'll try.

Reversibility

Math works like recipes. When we combine the correct ingredients in the proper order, we are rewarded with cookies.

In algebra, we often begin with a cookie, and try to figure out one or more of the missing ingredients. This works because of the principle of reversibility.

Reversibility says that anything that's legal to do in math can be undone by doing the opposite thing. If you double something, you can undo that by halving the result. If you multiply something by 10, then divide the result by 10 you'll wind up with the number you started with.

To construct a salad, we begin with tomatoes and cucumbers. My recipe looks like this:

"Cut up 1 tomato and 2 cucumbers into bite-sized chunks. Combine them in a bowl. Then eat it."

A more experienced cook could shorten the recipe. The first thing he'd do is eliminate references to eating it. That's outside the activity described by the recipe, which is creating salad in the first place. Using this same principle, he wouldn't tell you to go buy a tomato, or get up in the morning before you decide to cook. An experienced cook only wants the critical information. He knows what to do with it:

"1 tomato plus 2 cucumbers equals salad."

If a mathematician tried to create this recipe, she would not be able to resist shortening it even further by abbreviating:

$$1T + 2C = S$$

Adding is the opposite of subtraction. Reversibility says that if you subtract cucumbers from salad, you get tomatoes. If you subtract the tomatoes from your salad, you get cucumbers.

If you perform an operation on a number, then perform the opposite operation on the result, you'll wind up where you started. Start with the number 3, add 2 to it and you'll get 5. The opposite of adding 2 is subtracting 2. Subtract 2 from 5 and you'll be back where you started, at 3.

Reversibility is a god to whom mathematicians bring many bananas and mangos. This is not a small deal. Unquestioned faith in reversibility is Rule Number One in algebra. When the game gets a little complicated, when reality and algebra seem to disagree, we ignore reality.

The game must be kept internally consistent. This is not a bad thing. The whole thing wouldn't work without it. But it is a source of some confusion.

Because reversibility is so important, anything that won't bow down to it is excluded from the game. That's why we can't divide by zero, for example. Since multiplying anything times zero gives us an answer of zero, there would be no way to reverse dividing by zero. All our answers would be the same. We decide to keep reversibility, because it's very useful, and disallow dividing by zero. We can't have both.

The god of reversibility also dictates how shadowy creatures like negative numbers must behave. All of their activities must be reversible.

So how would negative tomatoes behave? Suppose you combine negative tomatoes with cucumbers and get salad:

$$(-T) + 2C = S$$

Reversibility commands that if you subtract negative tomatoes from your salad you'll get cucumbers:

$$S - (-T) = 2C$$

You might rebel at the idea of negative tomatoes. Can't eat one or throw it against a windshield. But, for the game to be consistent, if a negative tomato shows up, it will be subject to reversibility.

Perhaps you owe your neighbor a tomato. You might say you possess a negative tomato. You go to the store and buy one tomato and a couple of cucumbers, wanting a salad, but forgetting your tomato debt. Your refrigerator now contains:

$$1T + (-1T) + 2C$$

That is, one real, positive tomato, plus your to-mato debt, plus 2 cucumbers. When you come home, your cook has prepared dinner. If you see tomatoes in the salad, then the following has taken place:

$$1T +(-1T) +2C = S +(-1T)$$

You got your salad, but you still owe your neigh-bor a tomato.

If there are no tomatoes in the salad, perhaps the following happened:

$$1T +(-1T) +2C= 2C$$

The cook gave your tomato to the neighbor, and cut up your cucumbers for you. No salad.

Because you know the recipe, and you know the result, you can figure out what ingredients were used. Because you add the ingredients to get the salad, you sub-tract them from the result to figure out what's missing.

Multiplication is the opposite of division. If 3 times 4 equals 12, then 12 divided by 3 equals 4.

If $(3)(4) = 12$, then $\dfrac{12}{3} = 4$

It doesn't matter if you use numbers, or letters, or anything else. The game must be consistent. If you can prove, in the context of the game, that toads times gi-raffes equals elephants, then you must agree that elephants divided by toads equals giraffes, and that elephants di-vided by giraffes equals toads. If toads and giraffes are allowed to play the game, they're going to have to obey the rules.

Same with negative numbers, or imaginary numbers, or alien numbers. It really doesn't matter how you represent them. Reversibility rules! Within algebra, the truth is this:

$$\text{If } ab = c$$

$$\text{then } \frac{c}{b} = a$$

$$\text{and } \frac{c}{a} = b$$

Coefficients

"Coefficient" is a wonderful word. It sounds like nuclear physics, but represents the simplest idea in algebra. People will think you're a math jock just because you use it, yet you will master it in about 30 seconds. In fact, that's why it fools people. It's so simple, the books and teachers gloss over it quickly. But those of us who were sharpening pencils or pulling pigtails during that particular moment of our education may remain in awe of the word forever.

In this:

$$2x + 3A = 7$$

2 is the coefficient of x, and 3 is the coefficient of A. That's it. The coefficient is the number attached to an unknown, the number we multiply times the unknown. "7" is a constant, and not a coefficient, because there is no unknown next to it, no unknown being multiplied.

If we're talking about the real world, you might

see:

$$3 \text{ Rats} + 4 \text{ Snakes} = 2 \text{ Lawyers}$$

In this case, 3 is the coefficient of Rats, 4 is the coefficient of Snakes, and 2 is the coefficient of Lawyers.

   If an unknown, like "x," doesn't have a coefficient, we assume we've only got one of them. So, if you see:

$$x + x = y$$

you can assume the unstated coefficient of each of the x's and the y is 1. You could write it in without affecting the problem at all:

$$1x + 1x = 1y$$

   Coefficient is a word to cling to in confusing situations. You may not have the foggiest idea what to do when you see something like this:

$$33x^2 \bullet 12x - \sqrt{122} = y^2$$

But you can always raise your hand proudly and identify the coefficients. The coefficient of x squared is 33. The coefficient of x is 12. The unstated coefficient of y squared is 1. The square root of 122 is a constant, not a coefficient.

Parentheses

Soon after we learn to count, we are taught to add and subtract. Then we learn to multiply and divide. Taken together, these skills are called arithmetic.

Arithmetic may be annoying, but it rarely creates the terror and helplessness in its victims that the more muscular branches of mathematics inspire. This is because, on a conceptual level, we understand arithmetic. We might not remember what 7 times 8 produces, but it doesn't bother us. We can picture 7 boxes, each containing 8 rented videos we were supposed to return, and know intuitively that the late charges will be significant. At the worst, we can figure it on our calculators.

An *arithmetic* problem contains one type of operation, like addition.

An *algebra* problem, on the other hand, may contain several different arithmetic operations. Organization becomes important. We must keep track of several arithmetic problems. For example, we might express our problem in words like this:

"Multiply 10 times the sum of 10 and 40 and then double that answer."

This becomes tricky when we translate it into symbols. Perhaps we come up with something that looks like this:

$$10 \cdot 10 + 40 \cdot 2 = ???$$

The only inconvenience with this is that, if you're not careful, you'll get the wrong answer. For example, if you complete the operations from left to right, you'll be wrong, lost, and desperately confused. Ten times 10 equals 100, plus 40 equals 140, times 2 equals 280. Right?

Wrong. We don't do algebra from left to right. Instead, we obey the "order of operations." We multiply before we add. Someone seeing our problem, with no idea of our original intent, would multiply 10 times 10, then forty times 2, and then add the result. Their answer, 180, would be the right answer to what we've written, but not to our original word problem.

Obviously, we have failed to accurately translate our intentions.

Return to our original words. Multiply 10 times the sum of 10 and forty and then double that answer. We meant to take 10 times fifty (which is 500) and double that to get 1,000.

If we want to use the results of arithmetic problems we must organize them accordingly. We do this with parentheses:

$$((10 \cdot (10 + 40)) \cdot 2 = ????$$

We have now described the problem accurately. Parentheses serve as containers, boxes , for groups of items we want to treat as a unit. That's the key. We treat anything within parentheses as a single unit in the problem, as if it were a box of snakes. If your instructions are to carry the box across the room, don't reach in and carry reptiles one at a time.

In real life, we often instinctively group things together. We may ask the tenors to sing one part and the sopranos to sing a different one without feeling any panic at treating so many highly sensitive individuals as a single unit. We put all the household bills in one box marked "Due on the Fifteenth," and put the box on top of the refrigerator until about the 28th. Because this is so natural and common, sometimes we instinctively do it with math as well. But if we fail to group things correctly, we'll get the wrong answer. When the bridge we designed using our formula collapses, we'll all have a good laugh at our silly mistake.

We don't need to use parentheses to contain every item. We can say "2 plus 2 equals x" just fine:

$$2 + 2 = x$$

On the other hand, the meaning is not changed by packaging each character in a neat little box:

$$(2) + (2) = (x)$$

In fact, it's not changed by grouping the two's, either. Now we're describing a box which contains a box of 2 and another box of 2. This equals whatever's in the x box:

$$((2) + (2)) = (x)$$

Instinctively, in this case, you know what to do. You add 2 plus 2. "x" stands for "4."

People who've been fooling with algebra for some time instinctively group things, perform arithmetic operations on them in the correct order, and learn the correct answer. They may not be able to explain how they knew what to do.

Unfortunately, some of these people are math teachers.

A couple of concepts may help.

Multiplication is a game that only two can play. We group things in pairs any time we multiply or divide by more than one number. Then we complete the multiplications in a series of pairs:

$$2 \bullet 8 \bullet 15 = ??$$

In this case, it doesn't matter how we group them.

$$(2\bullet8) \bullet 15$$

gives us the same result as

$$2\bullet(8\bullet15)$$

The same is true of addition. It doesn't matter how we group an addition problem.

But some problems could have more than one meaning:

$$2 \bullet 8 + 3 \bullet 4 = ???$$

Do we add first, then multiply, or multiply first and then add, or what? Which of these numbers are grouped together? Our choices result in different answers:

$$2\bullet(8 + 3) \bullet 4 = 44$$
$$(2\bullet8) + (3\bullet4) = 28$$
$$((2\bullet8) + 3)\bullet4 = 76$$
$$2\bullet (8 + (3 \bullet 4)) = 40$$

They can't all be correct. Bad grouping gives us bad answers. Each of these might be the right way to describe some situation.

We treat the contents of parentheses as if they were a single unit. Any instruction placed next to the parentheses applies to the whole group. If there are no unknowns, we always have the option of completing whatever arithmetic is required within the parentheses. There's no change in the meaning of our algebraic expression if we change:

$$(\, 2 + 2 \,) \text{ to } (4)$$

Equally true, but less obvious, we can always substitute some equivalent item for whatever's within one of our little containers. We can change (4) into (2+2) without damage. Or, we could replace (16) with (4•4). Or with (2•2•2•2). There's been no change to the cargo within our container. They're all different ways of saying "16." Because we're treating the contents of the parentheses as a single number, a little internal shuffling doesn't affect the larger situation at all.

If we have been sloppy, we may wind up with a problem that looks like this:

$$(\, 4 \bullet 5 + 6 \,)$$

If we add 5 plus 6 first we get 11; multiply that times 4 and we get 44. On the other hand, if we multiply 4 times 5 we get 20; then we add the 6 and get 26. Which is correct?

The order of operations tells us. It says to do multiplication before we do addition, if the problem is unclear. Obviously, you're going to want to pay close attention to this "order of operations" business when you run into it.

Remember this: Parentheses, Exponents, Multiplication, Division, Addition, Subtraction.

If you can't remember that, remember "Please Excuse My Dear Aunt Sally." The first letters of the words will remind you of the order of operations.

Dr. Jim Calmly Discusses Algebra Books

The mathematics community insists on precise writing. But precision comes *after* you get the hang of things. You have to mess up quite a few times before the real elegance comes through, whether you're learning to ride a bike, bake cookies, or write a symphony. Math book writers would be less tense about their task if they "lied" a little while you're beginning to learn, and told you about the underlying ideas, instead of the precise details.

But publishers require them to tell the truth. So almost all math books are no fun to read and seem to be specifically designed to keep you from thinking. The great irony in this is that thinking carefully is what most mathematicians enjoy more than anything else. They love to figure things out and make absolutely sure that statements are always true. It's too bad they teach their courses and write their books so that you never get in on the excitement of the hunt and the intrigue of wondering how and why things work the way they do.

I bet that in the first and second grades you liked math activities. Most kids do, even today. What did you like the best? More importantly, when did it stop being fun? What happened? It may help if you remember that at one time you liked math. Perhaps you can again.

The good news is that many new curriculum reform efforts are designed to let students experience the frustration —and then the joy— that comes from working on and solving more interesting problems.

It's easy to imagine how math books came to be written in their present style. Guys who loved numbers, but weren't crazy about reading lots of words, brought their books to the print shop. The print shop guys were in the word business and thought all those complicated numbers were a waste of time. Between them, they managed to create a monster from the worst parts of each discipline.

Imagine the managers meeting in the Mathematics Section of the Gutenberg Publishing Company, hundreds of years ago.

"You've got to be kidding with this new algebra manuscript! Do you see how hard it is to typeset all these weird characters and display these equations. If we have any hope of competing with the Cheap Novels Division of the company, we have to figure out a way to cut our costs."

"I know!" said Wilhelm. "Let's just print every other line in the equations part. Eliminating half the stuff that the readers need to see will cut our work in half. Besides, we're doing people a favor by publishing this stuff at all. We have the right to expect our readers to work really hard to understand it."

"And what about all the helpful intuitive prose that our authors want to put in there?"

"Look, they want us to print those graphs, right? We can't do everything. Let's just edit out everything that helps it makes sense and print only the absolute minimum that needs to be there. Besides, who really needs to know math? Look where we got and we don't know any!"

"I have another great idea! If we only print every third line in all the equations, we'll save paper and make our books more exclusive."

"Great idea! But what about our competitors? What if they decide to put it all in? Won't we loose out in the market?"

"Did you forget? We don't have any competitors!"

And thus the modern era of mathematics publishing was born. Every succeeding generation of math authors and teachers read only books that were printed this way. Each generation passed its pain on to the next:

"Hey, we learned it from books like that. So can you!"

Algebra books contain long lists of exercises that often begin with a curt word or two that supposedly tells you what to do in each exercise. These words include commands like: substitute, evaluate or simplify.

Whenever you see one of them, you know you will not be doing interesting mathematical problems — just exercises. You will be practicing some particular isolated step that the book (and your teacher) feels is important to repeat over and over. I didn't like the exercises we had to do in P.E. classes. No football or basketball player ever does a push-up in order to make a touchdown or score a point. But football and basketball players know why they do push-ups and wind sprints. You should know why you are doing exercises too. Ask your teacher if you don't understand why each batch of exercises is important.

Many of these commands are clear. Some aren't. "Substitute" tells you to replace each letter with the numerical value that it stands for and figure out the numerical value of the expression. "Evaluate" means to find out what the expression equals, usually by a similar process.

The meaning of "simplify" depends on the context. It can mean to find out the numerical value of the expression. Other times it will mean to use the properties of the real number system, or combine similar terms and get a "simpler expression." Sometimes it refers to the steps used in solving an equation algebraically, as in "simplify this equation." When referring to expressions that have only numbers and no variables, "evaluate" and "simplify" are used interchangeably.

"Solve" means to find the numerical value(s) that make an equation true. "Solve: 5x - 9x = 12." They want you to find a number that will make the equation true when substituted for x. "Solve for x" means the same thing.

"Solve and check" means that after you solve an equation, you are to show your steps as you plug the answer into to the original equation and verify that it becomes a true numerical statement.

"Factor" means to write something as a multiplication problem. To factor 10 we might write 5 times 2. It really is that simple.

Beginning algebra jumps back and forth between learning to work with equations, and using number lines and graphs. They won't seem related because, at first, they're not. By the end of the year, they will be.

Fractions

We all were taught about fractions in elementary school, but we don't remember much about it.

The concept of fractions is simple: We have a pizza, we cut it into a certain number of slices. Everyone understands one-half of a pizza.

This familiarity contributes to our confusion. Manipulating fractions began as an activity that related to the real world. But, in order to keep the math game internally consistent, it developed moves that have no relationship to pizza, root beer, or life as we know it.

Fractions began by conveying informations like this:

$$\frac{one \; pizza}{three \; tenors}$$

This means we have one pizza, which we're going to divide among the three tenors. Each tenor will get less than a complete pizza. But each will get the same amount. The line can be read "divided by." That is, one pizza divided by three tenors.

Notice this: you aren't dividing pizza by pizza. That would be silly, wouldn't it? Nor are you dividing tenors by tenors. You can *add* three pizzas to your existing stockpile of pizzas just fine. You can *subtract* tenors from tenors and you wind up with a quieter group of singers. But it doesn't mean much to say three tenors *times* three tenors, does it? No. Multiplication and division often involve two different types of creatures.

It doesn't mean much to divide tenors by tenors, but we can divide them into groups. Six tenors can be divided into 3 duets, or into two trios. Sometimes we pay tenors not to sing. If we have 9 dollars, divided among the 3 tenors, each will receive 3 dollars. The silence will be a bargain. When we divide dollars by tenors, we create a fraction.

A fraction is a legitimate number in math. It has the same rights and obligations as any other number. It can be added to, subtracted from, multiplied by, divided by, etc. And any activity it engages in will be subject to the god of reversibility. You write a fraction as one number over another. Sometimes it's easier to read in a sentence if you use a slash mark between numerator and denominator. One-half would look like this: 1/2.

One fraction can have many names. Half a pizza is a certain amount of food. It doesn't matter how many slices we cut it into. If we are alone, we might make a single cut, slicing the pizza into two equal parts. This is easier to handle than a whole, uncut pizza for some of our daintier eaters.

If we cut that pizza into 20 slices, we could still identify half a pizza. Ten of these baby slices are exactly the same as one-half a pizza. If we cut it into six pieces, three of them equal half a pizza. One-half is simply a different name for five-tenths or ten-twentieths or three-sixths. Slicing it differently, or naming it differently doesn't change the amount of food.

Consider a whole pizza. This could be described as "1." It could also be described as 1/1, or one divided into one piece, which is to say, not sliced at all. Or it could be 8/8, or 8 slices of a pizza cut into 8 equal section. Or 123/123. All of those fractions simply mean "one." When we get into symbols, we might see x/x, and we will know this can be translated into a simple "1." If the numerator and denominator are the same, we're talking about "1."

It's easier to add fractions if we call them all by the same name.

What is 3 friends plus 5 paisanos? If you learn that "paisano" is Italian for "friend," it becomes simple. After translation, the question becomes: "3 friends plus 5 friends." The arithmetic is simple.

If we try to add 3 quartets to 5 duets, we might end up with several different answers. Each quartet and each duet could be considered an "act." If this is our translation, we end up with 8 acts. Perfectly good information. Or, we might translate quartet to mean "4 musicians," and duet to mean "2 musicians." In this case, we have 3 acts containing 4 persons and 5 acts containing 2 persons. We have 12 persons in quartets plus 10 persons in duets. We're going to have to pay 22 people for performing.

Every fraction can be thought of in two ways.

It's one thing (the numerator) divided by another thing (the denominator). Whenever you see 42/6 you are safe in reading it as "42 divided by 6."

You are also safe in visualizing many pizzas, each one sliced into six pieces, and our fraction describes 42 of these slices.

Multiplying Fractions

It's easy to multiply two fractions. You multiply the numerators (the top numbers) times each other, and the denominators (the bottom numbers) times each other. Your result will be a fraction, which you may choose to transform into some more manageable form. The reason you forgot that simple skill, acquired in fourth grade, is that it never made sense to you. Why does that work, anyway?

Any number times 1 equals itself.

Once you buy that rule, the god of reversibility says that any number divided by itself must equal 1. Ten divided by 10 equals 1. A million divided by a million equals 1. So, "one" can be stated as any number over itself:

$$\frac{giraffe}{giraffe} = one$$

$$\frac{sixty}{sixty} = one$$

$$\frac{xyz}{xyz} = one$$

Any fraction times 1 equals the original fraction. This simple problem can be written several ways. All are exactly the same. You choose a format to suit your purpose:

$$\frac{3}{4} \bullet 1 = \frac{3}{4}$$

$$\frac{3}{4}(1) = \frac{3}{4}$$

$$\frac{3}{4}\left(\frac{1}{1}\right) = \frac{3}{4}$$

In fact, you might choose to describe "1" the way we just talked about. One is any number over itself. Any fraction in which the numerator and denominator are the same. So:

$$\frac{3}{4} \bullet \frac{569}{569} = \frac{3}{4}$$

If we examine the result of multiplying any fraction by any variety of "one," we realize that we multiplied numerators times each other, and denominators times each other.

It works just as well for fractions other than "1."

$$\frac{2}{3} \bullet \frac{7}{8} = \frac{14}{24}$$

We multiplied 2 times 7 to get 14, and 3 times 8 to get 24. To make sense of this, replace the multiplication sign with the word "of." Instead of "One-half times one-half pizza," think of "One-half of one-half pizza." which is one fourth of a pizza.

That's all there is to multiplying fractions. Multiply the first numerator times the second numerator to get the answer's numerator. Then multiply denominators times each other to get your answer's denominator.

Here's a geometric model for what is happening in fraction multiplication. The problem 2 • 5 = 10 is really about how many tiles it takes to cover the bathroom floor:

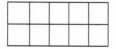

This same model works for multiplying fractions. Here's a square room covered with floor tile:

To depict the multiplication problem

$$\frac{2}{3} \cdot \frac{1}{6} = ??$$

we first divide the room horizontally into three slices and shade two-thirds of it:

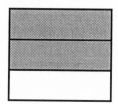

Repeat the same idea vertically for the fraction 1/6:

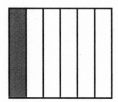

If we overlay these two shadings, we can see the product $\dfrac{2}{3} \cdot \dfrac{1}{6} = \dfrac{2}{18}$

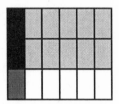

The whole tile is divided into eighteen (3 • 6) sections and two sections (1 • 2) out of those are shaded.

Here's what $\dfrac{3}{4} \cdot \dfrac{2}{3}$ would look like. The answer is $\dfrac{6}{12}$ or $\dfrac{1}{2}$.

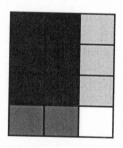

Dividing Fractions

To divide one fraction by another, invert the second one (put the bottom number on top and the top number on the bottom) then multiply numerator times numerator and denominator times denominator. This is a way of "reversing" the multiplication process.

What in the world does it mean to divide one fraction by another fraction? If division is the opposite of multiplication, does that mean we're stitching pizza slices back together?

$\dfrac{12}{3}$ asks how many three's there are in 12. Similarly, $\dfrac{2}{3} \div \dfrac{1}{9}$ asks how many one-ninths of a pizza there are in two-thirds of a pizza. There are six. A problem like $\dfrac{1}{6} \div \dfrac{1}{3}$ asks how many thirds there are in a sixth. Not even one. So the answer is a fraction.

Perhaps the orchestra conductor intends to pass out new music at tonight's rehearsal. She ordered 60 copies, 1 for each member of the orchestra. But only 10 copies arrived. She's going to have to divide the music like this:

$$\dfrac{1}{6} \text{ music} \div \text{orchestra members}$$

That night, two-thirds of the orchestra calls in sick. So she's going to distribute the music she has among one—third of the total orchestra members.

$$\frac{1}{6} \text{ music} \div \frac{1}{3} \text{ orchestra members}$$

How many members will receive music?

If $1/3 \cdot 1/2 = 1/6$, reversibility tells us that 1/6 divided by 1/3 equals 1/2 . (It also tells us that 1/6 divided by 1/2 equals 1/3.)   We know what the answers are. She had 10 sheets of music, and divided them among 20 orchestra members. Only half the players received music. Now we just have to figure out what we did. One way we could have gotten:

$$\frac{1}{6} \div \frac{1}{3} = \frac{1}{2}$$

Is by taking the second fraction, inverting the numerator and denominator, then multiplying the two fractions together:

$\frac{1}{6} \div \frac{1}{3}$ changes to $\frac{1}{6} \cdot \frac{3}{1}$ which equals $\frac{3}{6}$ which is simply

another way to say $\frac{1}{2}$.

To divide one fraction by another, invert the second one (put the bottom number on top and the top number on the bottom) then multiply numerator times numerator and denominator times denominator.

Adding Fractions

The first step in adding fractions is to make sure the denominators are the same. If I cut my pizza into 5 pieces, and you cut yours into 8 pieces, and I eat 2 pieces of mine and 6 of yours, will you get as much as me?

First we write out the situation. Each of my pieces is one-fifth of a pizza. I ate 2 of them. The amount I ate looks like this:

$$\frac{2}{5}$$

Each piece of your pizza is one-eighth of a whole pizza. I ate 6 of them. The amount I ate looks like this:

$$\frac{6}{8}$$

To see how much I ate altogether, we add these two fractions together:

$$\frac{2}{5} + \frac{6}{8} = ??$$

But now we're trying to add apples to oranges again, so to speak. The slices are different sizes. Merely telling someone I ate eight slices won't describe how full I should feel. We need to translate, so that our denominators (the bottom numbers) are the same. We need to identify some common denominator.

We might try to create a chart that included all the different ways to express the idea of 2/5, then make a chart showing all the ways of saying 6/8. Hopefully, somewhere on this chart there will be two numbers that have a common denominator.

To create a chart, we multiply each of the fractions by all the variations of the number "one" we can think of, because we know that it won't affect the size of the fraction at all. Anything times one equals itself:

$$\frac{2}{5} \cdot \frac{1}{1} = \frac{2}{5} \qquad\qquad \frac{2}{5} \cdot \frac{3}{3} = \frac{6}{15}$$

$$\frac{2}{5} \cdot \frac{2}{2} = \frac{4}{10} \qquad\qquad \frac{2}{5} \cdot \frac{4}{4} = \frac{8}{20}$$

And so on.

Each answer is simply a different name for the same amount of pizza.

Using the same technique we discover that 6/8 is the same as 9/12, and 12/16, and 15/20, and 18/36, and 21/28, and 24/32, and 30/40, and 36/48, etc. Each of these is a different name for 6/8.

Both charts have numbers that include denominators of 20. They also include denominators of 40. Either one will work. If we continued our chart, we could add many more "common denominators." They'll all work just fine. We see that we are really adding 16/40 (for example) plus 30/40. Our answer is 46/40. Obviously, I ate just a tad more than my share. But then, I never claimed to be good at this math stuff.

. In ancient times, (before hand-held calculators) math teachers spent a lot of time torturing students with the concept of "least common denominator." This means the smallest common denominator. In our little example, they would have insisted we use "20" as our denominator. When calculations were performed with a pencil on a Big Chief tablet, or a stick on a wax board, or charcoal on the cave wall, using the smallest common denominator meant smaller numbers and less chance of error. These days, with the aid of calculators and computers, children can multiply and divide six digit numbers before they have dependable bladder control. Since any common denominator works as well as any other, why risk the brain damage of seeking the smallest one? There's always one easy one. If we multiply the denominators times each other, we end up with a number that is guaranteed to be "common." Can't help but be. To add 3/4 plus 4/5, since four times five equals twenty, we know twenty is one of the common denominators. If we translate both fractions into twentieths, we can just add them up. That leaves the question of how to determine the numerator, or top number of each side.

$$\frac{3}{4} + \frac{4}{5} = \frac{?}{20} + \frac{?}{20}$$

We could do this by creating a chart, multiplying 3/4 by other translations of the number one, like we did before:

$$\frac{3}{4} \bullet \frac{1}{1} = \frac{3}{4} \qquad\qquad \frac{3}{4} \bullet \frac{4}{4} = \frac{12}{16}$$

$$\frac{3}{4} \bullet \frac{2}{2} = \frac{6}{8} \qquad\qquad \frac{3}{4} \bullet \frac{5}{5} = \frac{15}{20}$$

$$\frac{3}{4} \bullet \frac{3}{3} = \frac{9}{12} \qquad\qquad \frac{3}{4} \bullet \frac{6}{6} = \frac{18}{24}$$

Three-fourths, translated into twentieths is fifteen twentieths. A chart of 4/5 would look like this:

$$\frac{4}{5} \bullet \frac{1}{1} = \frac{4}{5} \qquad\qquad \frac{4}{5} \bullet \frac{4}{4} = \frac{16}{20}$$

$$\frac{4}{5} \bullet \frac{2}{2} = \frac{8}{10} \qquad\qquad \frac{4}{5} \bullet \frac{5}{5} = \frac{20}{25}$$

$$\frac{4}{5} \bullet \frac{3}{3} = \frac{12}{15} \qquad\qquad \frac{4}{5} \bullet \frac{6}{6} = \frac{18}{24}$$

Now we can finish our translation.

3/4 +4/5 is the same as 15/20 + 16/20

Fifteen slices plus 16 slices equals 31 slices. The answer is 31/20.

Perhaps we don't want to invest the time to create a little chart every time we need to find the right numerator, after we've identified a common denominator. Another little trick can save us time. Look what happened in our last problem. We learned that:

3/4 +4/5 is the same as 15/20 + 16/20

We found one common denominator by multiplying the two denominators by each other. Four times five equals twenty, we used twenty as the common denominator. We really multiplied 3/4 by one, but chose the "one" built on the other fraction's denominator.

$$\frac{5}{5}$$

And we multiplied the 4/5 by one, but chose the variety of "one" built on the other fraction's denominator:

$$\frac{4}{4}$$

So, to find the numerator, we also have to multiply each *numerator* by the other fraction's *denominator*. To multiply a fraction by 4/4, we multiply both numerator and denominator by 4. That's what we're doing here. And, of course, we're really multiplying by a variety of "one."

Once the logic sinks in, translating fractions into equivalent fractions with common denominators will be child's play.

Subtracting Fractions

Subtracting fractions is the opposite of adding them. No inverting things, no fancy rules. First you make sure you have common denominators, then subtract the second numerator from the first one:

$$\frac{45}{63} - \frac{42}{63} = \frac{3}{63}$$

Reducing Fractions

We can make fractions easier to work with by translating them into smaller numbers. One million over 2 million is awkward. The ratio is simple, however. It's one to two, or 1/2.

We don't change a fraction at all by multiplying both the numerator and the denominator by the same number. We just translate it.

Reversibility commands that the opposite shall also be true. We don't change a fraction if we divide both the top and bottom by the same number. We call it reducing, but it's really translating. You'll feel a lot thinner if you stop thinking about how much you weigh in grams, and start thinking in terms of pounds. In fact, you probably weigh less than a tenth of a ton. Changing your denominator is a lot easier than dieting.

If we see the fraction 10/20, we can multiply both numerator and denominator by 10, and the answer,

$$\frac{100}{200}$$

refers to the same amount of pizza. Reversing this, we can divide both numerator and denominator by 10 and get back to 10/20. If we divide both these by 10 again, we'll get 1/2. Again, we haven't changed the fraction. We're just calling it by a different name.

In real life, there's a big difference between half a pizza and 100 slices of a pizza that's been cut into 200 tiny slivers. Once again, for the purposes of the game, forget reality.

Just as you can multiply both numerator and denominator by the same number, you can also divide them both by the same number.

A Little Note

If we can multiply both numerator and denominator by the same number, without changing the fraction, can we also add or subtract the same way?

No. We can't. Common sense tells us this won't work. If we start with 1/2, for example, and try to add 1 to both numerator and denominator, we wind up with 2/3. Two-thirds is not the same as one-half. You can multiply or divide both numerator and denominator of a fraction by the same number, and nothing significant is changed. But you can't add or subtract.

Whole Numbers Interacting with Fractions

Any number divided by 1 equals itself. Any number multiplied by 1 equals itself. So, you can represent any number as itself over 1.

$$43 = \frac{43}{1}$$

Now you have a fraction. If you put it next to another fraction, they'll behave just like two fractions are supposed to:

$$\frac{43}{1} \bullet \frac{14}{16} = \frac{602}{16}$$

We could divide 602 by 16 and use that result as the answer.

Sometimes we run into a number like six and two-thirds. That's really 6/1 +2/3. In order to use it as a single fraction, we have to add these two together. That means converting them to a common denominator. If we multiply 6/1 by 3/3 we get 18/3. Add this to our 2/3 and we get 20/3, unfairly called an "improper" fraction because it's greater than "1."

Terms

The basic "game piece" of algebra is the "term." Any individual number or letter is a term. Some groups of things are also considered "terms." Anything we add or subtract is called a term. In the equation

$$2 + 2 = 4$$

each "2" is a term, and so is the 4. In the equation

$$435 + 64 + \frac{1}{34} + \text{Rhino} - \text{Robin} = \text{Mass Confusion}$$

there are six terms. The fraction is a term, the numbers you add are each terms, the Rhino, and the bird you wish to subtract are all terms. "Mass Confusion" is a term, as well as your mental state.

This is so simple, your textbook will say it in about three words, and you'll forget it. Your instructor will explain it quickly, probably while you're trying to finish the essay that's due next period in English Literature.

Later, trying to decipher your notes, you may doubt that it's this simple. But it is. Anything you add or subtract is called a term.

In algebra, terms can be constructed to look frightening indeed:

$$\sqrt{678} + x^{35} - xyz^2 + \frac{3}{8} = \text{Yellow Sub} + \text{Walrus}$$

There are four terms to the left of the equal sign, and two to its right, each separated by either a plus sign or a minus sign.

Factors

If we're adding and subtracting, we talk about terms. When we multiply, we talk about factors and products. When we multiply tangerines times bananas, both bananas and tangerines are "factors." We could talk about a problem like this:

$$(x + 1)(\text{zebras}) = \text{Math Whiz}$$

"x plus 1" is one factor, "zebras" is another factor, and "Math Whiz" is their product. In the problem:

$$8x \cdot 10 = 80x$$

"8x" is one factor, "10" is another factor, and "80x" is their product.

Kenn's Polynomial Dilemma

Many people love the symmetry and precision of language. The rich history of English helps them understand and remember unfamiliar words. They know an "octogenarian" is an eighty-year-old person, because "octo" always means eight, and "genarian" always refers to someone's generation or age. People blessed with this sort of mind usually learn to play complex stringed instruments, like violin or guitar.*

These "word guys" can expect to be swimming in the bilge water while they try to corral the concept of "polynomials," which we'll talk about in the next chapter. The problem is this: over the last fifty years, textbooks have explained the word "polynomial" inconsistently. Today's math teachers learned several slightly different versions. Because math teachers are usually "number guys," this doesn't bother them. Depending on how your teacher learned, and how pig-headed you are, it may bother you.

Word guys recognize the prefix "poly" as meaning "many." A polygon is many sided, and polymorphous means having many shapes or forms. Language-oriented folks see the word polynomial and say, "many what?" Some teachers and books reply, "many terms." A polynomial is an expression with more than one term. Later, the book may say that something with only one term is also a polynomial. All the guitar players immediately dive off the truck.

Other books say that every whole number, every unknown and every combination of them is a polynomial. That means the number "1" is a polynomial. All the violin players scream in unison.

Editor's Note: Kenn plays guitar.

74

Other books say that "1" is certainly not a polynomial. A one-term expression can't be a polynomial.

The solution was obvious. I had to create my own definition that was consistent with how all the math teachers actually *use* polynomials, while maintaining the integrity of the language. It was the only way I could get any sleep.

Your teacher may disagree with some details, but she won't hassle you, because she wants you to solve problems correctly, not define words. Shoot, Jim doesn't quite agree with me, either. The only reason he's letting me include my own little world-view is that it allowed me to accept polynomials into my life. Until I reconciled the word with a consistent meaning, I just couldn't grasp the concept. There may be other students as stubborn as me who will find this useful. And, as Jim points out, there may be teachers out there who have no idea this is a problem for us.

Polynomials

In arithmetic, we worked with numbers. In algebra, we work with numbers as well as variables and unknowns. A mixed group of numbers and unknowns that we treat as a single unit, (like "35x") is called a "polynomial." If "2x + 33" is treated like a single group, we call the entire group a polynomial. In my little mythology, in algebra we *always* work with polynomials. Numbers, once transported into the game, act like polynomials. They have more than one component, the number component and the unknown part.

Even a simple integer like "14" can be considered a polynomial, because it's the same as "$14x^0$" And a single unknown, like "x" can be considered a polynomial, because it's the same as "$1x$."

Of course, there's no big advantage to giving fancy new names to old, familiar numbers and unknowns. The word "polynomial" is usually used for game pieces consisting of sums, differences, and products of numbers and letters. Each of these groups is a polynomial:

$$3x^2 + 2x - 43 \qquad 16x^4 + 64 \qquad x + 3 \qquad 127x^5$$

We label polynomials according to how many "terms" they have. You will know something is a "term" because either it stands alone (like "23" or "2x") or you're asked to add or subtract something to it within its group. In a situation like:

$$23 \cdot 16 = x$$

our factors are 23 and 16. Simple numbers. No internal adding or subtracting going on with either of them. Each of them is a "monomial," which means a "1 term" polynomial. Notice this: A monomial, having only one term, is a variety of polynomial. Your teacher may disagree, depending on what book she had in college. Some people feel strongly that the word "polynomial" does not include monomials. They say we have one term expressions (monomials) and expressions with more than one term (polynomials). Other folks say a monomial is a one term polynomial.

We don't feel particulary strongly either way. If the text book writers are confused, it's reasonable for you to be a little confused too. Accept your teacher's definition. As a practical matter, when you see the word polynomial, you'll be working with monomials and expressions with more than one term.

Even fancier items can be monomials:

$$xyz \bullet 2x \bullet 3y = \text{root rot}$$

None of our characters displays any internal adding or subtracting. They are all monomials, although because of the context, they'd be called factors here. Compare that to this:

$$(x + y) \bullet 32 = z$$

We're multiplying the result of x + y times 32. We treat (x + y) as a unit because it's inside the parentheses. Because "x + y" has two terms, we call it a "binomial." We know it has two terms, because we see that plus sign between them. A binomial is a polynomial with two terms. Three-term polynomials are "trinomials."

This simple concept drives people crazy. Are there any polynomials in this equation?

(3 tenors + 5 basses)(3 solos + 6 flags) = (eggs)sopranos

Yes there are. We have two polynomials to the left of the equal sign. What variety of polynomial are they? They are each binomials, because they have two terms. The stuff to the right of the equal sign is a monomial, because it contains no internal adding or subtracting. We might think of it as an "expression," or as a "member," or as a monomial, depending on what we intend to do with it. The definitions refer not only to what something "is" but also what role it's playing.

A "trinomial" might look like this:
(3 tenors + 4 basses + 7 altos)

The Distributive Principle

Although we treat the contents of parentheses as a single unit when we're translating real life into an equation, we have several options of how to deal with those contents. There may be strategic advantages to choosing one way over another.

The distributive principle is common sense. It tells us that if we're going to multiply something times each number within a group, and then add them all together, it doesn't matter if we add them all up first, and then multiply, or multiply times each member of the group and add the results.

Perhaps we started with tenors in parentheses, like this:

(tenors)

When we decide to flog the tenors it might look like this:

Flog (tenors)

Alas, perhaps we are not equipped to flog more than 1 tenor at a time. We list our tenors within the parentheses:

Flog (Bill+Dwight+Kevin)

Doesn't change the meaning. To conduct the actual flogging, we can focus our attention on each in turn:

Flog Bill + Flog Dwight + Flog Kevin

Similarly, we might decide to kiss altos:

Altos = Jane + Sally + Paula

Therefore,
Kiss (Altos) =
    Kiss (Jane + Sally + Paula) =
        Kiss Jane + Kiss Sally + Kiss Paula.

It doesn't matter which order you do your kissing or flogging. When you complete your operations, you'll have three sore tenors and three blushing altos.

Perhaps our parentheses contains 3 quarters, 2 dimes, and 40 pennies. We bet the whole piggy bank on an ice hockey game and win. We double our money:

double (3 quarters + 2 dimes + 40 pennies)

We can count our money and double it. Or we can double our quarters, then our dimes, then our pennies, then add them all up.

Let's substitute Q for quarter, D for dime, and P for penny. The word "double" really means "2 times." We can write out our situations like this:

$$2(3Q + 2D + 40P)$$

Algebra has taken something simple and lovely, like betting on hockey games or kissing altos and made it look scary.

But, the truth is, you already know what to do with this. Because every Q represents 25 cents, every D represents 10 cents and P equals a penny, you could replace the letters with the appropriate numbers, complete the math within the parentheses, and then multiply the answer times 2:

$$2\big((3 \cdot 25) + (2 \cdot 10) + (40 \cdot 1)\big)$$

which equals:

$$2(75+20+40)$$

which equals

$$2(135)$$

which equals 270.

Or, you could simplify the equation by multiplying 2 times each number in the parenthesis.

$$2(3Q +2D + 40P)= 6Q+4D+80P$$

This is just as simple, though not as fun, as flogging tenors.

Perhaps it wasn't a simple hockey bet. Perhaps it was a pool. Several people bet the contents of their piggy banks, and you have just won it all. They each contributed 3 quarters, 2 dimes, and 40 pennies:

$$(3Q +2D + 40P)$$

Now if you win, you don't merely double your winnings. If five people bet, you make five times your investment. If ten people were in the pool, you make ten times your investment. Before your bookie tells you how many people contributed, you can use an x to represent that number. Your winnings will be:

$$x(3Q +2D + 40P)$$

Treat x the same as you treated 2 earlier:

$$(x \cdot 3Q)+(x \cdot 2D)+(x \cdot 40P)$$

Within the parentheses, we only have multiplication problems. It doesn't matter what order you multiply a series of things if that's all you're doing to them. We can make it look neater by moving the x's.

$$3Qx + 2Dx +40Px$$

This is a tradition in algebra: Put numbers first, followed by unknowns. 3Qx means the same things as x3Q or Qx3. But people have become so accustomed to seeing things one way they become confused if you change the tradition. They also neglect to tell you about it.

When our bookie tells us how many of our friends contributed to the pool, we'll substitute that number for the x's, do the multiplication, then the addition, and have our answer.

Imagine a chocolate bar that has all of those sections filled with sweet raspberry juice. It is 3 sections by 8 sections when we buy it.

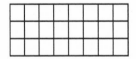

There are 3 • 8 = 24 sections. Now if I break off 2 rows you'd see this for a brief moment.

Not only is this chocolate bar extremely tasty, it is a perfect illustration of the distributive principle.

$$24 = 3 \cdot 8 = 3(2 + 6) = 3 \cdot 2 + 3 \cdot 6 = 6 + 18 = 24$$

In your Real Algebra Book it looks like this:

$$a(b + c) = ab + ac.$$

As further evidence of their senses of humor, text-book writers will refer to the distributive law, then the distributive principle, then the distributive property. These are simply different names for the same beast. Writers play the same game with the commutative and associative whatchamacallits we are about to investigate.

===

The Commutative Principle

It doesn't matter what order you add or multiply things, if that's all you're doing. That's the commutative law (or principle, or property.) It comes from the same Latin word that gave us "commuter." You drive to work, you drive home from work, and your car shows the same distance traveled each way.

When adding a long series of numbers, you can start at the bottom, or the top, or the middle. If you've got a bunch of numbers to multiply times each other, and that's all you want to do with them, it doesn't matter what order you do them in. Flog your tenors in whatever order pleases you most. They'll all be singing the same song by the time you're done.

Algebra holds the commutative principle very dear. It always works, and we use it all the time. We re-arrange numbers in an equation to make them easier to deal with, safe in our knowledge that it doesn't matter what order we add things up, and it doesn't matter what order we multiply a list of items, as long as that's all we're doing to them.

But if you're making a graph, or a map, it does matter. You drive five miles north, then three miles east to your girlfriend's house. If you mistakenly drive three miles north, then five miles east, you're in the middle of a cornfield.

In the 1800s, a fellow named Hamilton imagined a kind of math that was not commutative. Where it did matter what order you multiplied things out. Where 3 times 5 was different than 5 times 3.

No one paid much attention to Hamiltonian math, until Einstein and the boys started trying to visualize quantum physics. Regular algebra couldn't represent what they were observing. Hamiltonian math did the trick.

You should forget you ever heard this, for three or four years. Instead, you should memorize the commutative principle, which says that it doesn't matter what order you perform multiplication or addition.

The Associative Principle

Organizing your singers into sections won't affect how many are in your choir. Grouping them is creating associations. The associative law recognizes the benefits of confining your tenors to one easy-to-patrol area. Fifteen tenors in a corner is no different from fifteen scattered throughout the room. It's just safer.

If all you're doing to a series of numbers is adding them, you can add them one at a time, or you can put them in groups, then add the totals of each group. 10 plus 3 plus 5 is the same as (10 plus 3) plus 5, or any other combination. Sometimes you'll discover an advantage to creating groups. Other times, things will come to you organized one way when it would be more convenient to group them a different way, or to eliminate the groupings altogether. The associative principle gives you permission to do that.

The easy way to group things is to put them inside parentheses. If all you're doing is multiplying or adding a series of numbers, but not both, it doesn't matter how you group things within parentheses.

Notice that there is no similar "law" for subtraction or division. It does matter what order you do those operations, so you have less freedom to be creative with grouping them, or associating them, together.

The Little Weenie Numbers Approach

When confronted with a mathematical situation that gives you the creeps, you may consider a time-honored practice that is seldom described outside the sacred smoky kivas in the basements of mathematics buildings:

Try Little Weenie Numbers.

Equations with fancy symbols learn to subdue their prey with attitude as much as with fang and claw. One defense is to construct equations similar to your attacker, but with Little Weenie Numbers, like "2," or "3," or "5."

The arithmetic will be quicker, and the results will be harmless little puppy numbers.

The Little Weenie Approach won't give you the answer to a problem. What it will do is let you experiment with different techniques. This is especially useful if you've forgotten how something is supposed to work. Say you've got to multiply a couple of polynomials:

$$(a+b)(a-b)$$

You know this is a common formula, you know you were supposed to memorize how the result would look. But aliens have confiscated that portion of your brain, and you need to figure it out right now, during the test. In fact, instead of "a" the teacher may have substituted some howling, raging decimal, or a seething, rabid fraction, and you don't have time to try several different calculations.

So, to help remember how the game is played in this situation, you might substitute "3" for "a" and "2" for "b". Now you can try whatever options are confusing you, and see quick results. You know what the correct answer will be, because you can conduct the arithmetic so quickly. 3 plus 2 is 5. 3 minus 2 is 1. 5 times 1 is 5. The answer is 5. So you might try, for example, multiplying "a" times each number in the second factor, then multiply "b" times each number in the second factor, then combine like terms. If that process also gives you the answer of "5" you've got good evidence that your strategy worked.

A word of caution: your teacher may not approve of the Little Weenie Numbers Approach. It empowers you to check yourself (and him) and to confirm your understanding of various strategies. It also means that you can face down dangerous equations without a text or an abundantly-degreed bodyguard. This could threaten your teacher's livelihood. In an emergency, you might even learn whole new concepts without paying for protection services.

Clearly, this is not good for anyone but you. Don't tell them where you got the idea.

Canceling Parts of Fractions

As we watch experienced math show-offs speed through problems, we notice they often do an odd thing. When they see a fraction with the same number on the top and the bottom, they simply cross them both out.

When you or I try this at home, we often get the wrong answer. What is the secret to this magical little shortcut?

We can reorganize fractions into shapes that are more convenient to work with, so long as we're careful not to affect how much pizza they represent. We can describe $\frac{3}{4}$ as $\frac{6}{8}$ for example, if we choose. We could also describe

$$\frac{apes + baboons}{chimps}$$

just as accurately this way:

$$\frac{baboons + apes}{chimps}$$

When we rearrange fractions in that way, we can move terms around at will. It doesn't matter what order you add or subtract, if that's all you're doing. But you can't separate factors from each other so casually. The terms "15x" means 15 times x, and that's completely different from baboons times x. So, when you rearrange a fraction to make it more understandable, keep factors joined. So, you could rearrange:

$$\frac{13(pear\ tree) + basketcase + tree}{tree + 13(pear\ tree) + basketcase}$$

to look like this, and you won't have changed the fraction at all:

$$\frac{13(pear\ tree) + basketcase + tree}{13(pear\ tree) + basketcase + tree}$$

Once reorganized like this, you may notice a pattern within the fraction. In fact, in this case, you now know that this fraction is equal to "1."

You can't split the 13 from its pear tree without screwing things up. You might have foolishly come up with something that looked like this:

$$\frac{(pear\ tree) + 13basketcase + tree}{13(pear\ tree) + basketcase + tree}$$

Does that equal 1? I think not. Thirteen times basketcase is much different than 13 times pear tree. Ten times red car is not the same as 10 times horse droppings. The point is, when you rearrange fractions for convenience, things you add or subtract can be moved at will, but things you multiply must stay together.

Hold that thought.

When we multiply anything by 1, the answer is the original number. One times 40 is 40. One times a billion equals a billion. Multiplying by one doesn't change the original number.

This is obviously different than adding 1. Every time we add 1, we change the original number. Forty plus 1 equals 41, not 40. A billion plus 1 equals a billion 1, not merely a billion.

This is the secret behind the little "cancel the number" trick. A number in the numerator cancels a number in the denominator if the two of them really represent multiplying the rest of the fraction by "1."

If we see:

$$\frac{chimps + baboons + apes}{apes(monkeys) + chimps}$$

we notice apes on both top and bottom, and also chimps on both top and bottom. Can we cancel out one or both of these?

It would be easier if we reorganized the problem:

$$\frac{apes + baboons + chimps}{apes(monkeys) + chimps}$$

Our highly trained mathematical eyes leap to "chimps over chimps." And because of keen insight, refined memory, and because this book has hammered it into us, we know that chimps over chimps equals "1." We are *adding* "1" to both numerator and denominator, not *multiplying* them. We can't cancel chimps.

But how about "apes?" In our denominator, we're multiplying apes times monkeys. Can we cancel both apes out?

The short answer is, "no."

Perhaps this would be a good time to test our "Try Little Weenie Numbers" concept. Take the basic design of our primate problem above and substitute Little Weanie Numbers:

$$\frac{6+3+4}{6(5)+4} = ???$$

We can figure out what this fraction equals by doing the arithmetic. 6 plus 3 plus 4 equals 13. 6 times 5 equals thirty, plus 4 equals 34. This fraction is the same as:

$$\frac{13}{34}$$

If we try to cancel out the sixes, we get seven over nine. 13 thirty-fourths is less than half, while seven-ninths is more than half. They can't be equal.

So the Little Weenie Number Experiment proves that you can't cancel numbers if either one of them is occupied with addition or subtraction.

The only time you can cancel is when you are really *multiplying the rest of the fraction by 1*. If a number is busy multiplying the rest of the numerator, and that same number is busy multiplying the rest of the denominator, you're multiplying the fraction times 1. In that situation, you can just cross both of them out and move on.

$$\frac{10(x+y)}{10(z+y)} = \frac{x+y}{z+y}$$

In this case, both the 10 on top and the 10 on the bottom are doing just that. They are prime candidates for cancellation. We cross them out and pretend they never existed.

The important thing to remember is that a critter you can cancel has three bright markings:

It has to be in both numerator and denominator. That is, it has to show up on both the top of the fraction and the bottom.

It has to be involved in multiplication, not addition or subtraction.

More than that, it can't be multiplying only a portion of the numerator and denominator. That wouldn't be the same as multiplying the whole thing by 1. It has to be multiplying times all the rest of the numerator and denominator.

═══════════════════════════

An Exception to Canceling

Teachers don't like to confuse beginning students with the truth. Often they prefer to confuse you later, when you're no longer a beginner.

When they talk about canceling fractions, early in the class, they will say that you can cancel unknowns, and polynomials just as easily as you can Little Weenie Numbers. Happy and gullible as you are, you'll skip along with them.

But there's a danger lurking along that pathway.

An unknown could be any number. If you plug certain numbers into an equation, you're going to discover yourself dividing by zero. Or multiplying the whole works by zero and annihilating your entire equation.

In many cases, you can treat a polynomial like (x+3) just like any other suspect for canceling. You'll cancel them in numerator and denominator, come up with the correct answer, get a good grade, and look forward to class.

When you get into calculus, they won't let you do that any more, willy-nilly, whenever the mood strikes you. Factors containing an unknown will have to be canceled with special care.

It just seemed fair to warn you.

Exponents

Multiplying something times itself happens so often, they've given the process its own little world, complete with rules and shortcuts.

The number we're going to multiply times itself is called the "base."

The number of times it's multiplied by itself is determined by the "exponent," a little number raised a little and hung up to dry beside the base.

If you multiply a number times itself, you say you've squared it, and you represent that by hanging a little 2 up next to the base.

Sheep^2

means "sheep" times "sheep" and you pronounce it "sheep squared." Sheep is the base, 2 is the exponent. They use the word "square" because you find the area of a square this same way, by multiplying the length of a side times itself.

$$(sheep)(sheep) = sheep^2$$

If you multiplied rats times rats times rats, like this:

$$(rats)(rats)(rats)$$

you could express the same thing this way:

$$(rats)^3$$

which is pronounced "rats cubed." In this example, you used "rats" as a factor three times. You could also say you "raised rats to the third power." The word "power" refers to exponents. Squaring is raising to the second power. An exponent of 12 is the same as "raising to the twelfth power."

Mathematicians' constant struggle to achieve consistency within their game has led to some interesting quirks regarding exponents.

At some ancient scholarly party, a tipsy mathematician proposed a puzzle. If you line up three identical items and multiply them, you have cubed the item, or raised it to the third power. The exponent will be a 3. If you line up two identical items and multiply them, you have squared the item, and the exponent will be a 2.

So, what happens when you line up only 1 item? You've got an exponent of 1, but no multiplication to do. But you do have that one item, lined up all by itself. The consistent, fair thing to do is to declare that raising something to the first power is the same as doing nothing at all.

$$Porcupines^1 = Porcupines$$

As the conversation flowed ever more freely, they got a little more outrageous. What would it mean to use zero as the exponent?

What does x^0 stand for?

Your first, healthy reaction, is to declare this an absurd concept. What in the world can it "mean" to raise something to the zero power? Squaring something is multiplying it times itself. Raising it to the first power means doing nothing at all. How can anyone, except perhaps a tenor, do even less than that?

If the idea of using zero as an exponent seems twisted and bizarre to you, that clearly indicates that you don't know how to have a good time like these ancient mathematicians did.

But the idea is there, a base raised to the zero power, and perhaps it has a meaning that is consistent within the algebra game. If so, we may be able to employ it, even though it is obviously a ridiculous concept, with absolutely no relationship to reality. That's the leap your mind must make. Accept the idea that we may have game pieces that work just fine to turn the gears and spin the wheels of the algebra engine, but that make no sense at all when extracted and examined in the cold light of logic.

Any number, raised to the zero power, equals 1.

No logical person would guess that. But this was the only choice that kept the rest of the game consistent because of how exponents work when we use them in equations.

By using the Little Weenie Numbers approach, guys no smarter than you or me started noticing some patterns. For example, what happens when you multiply 3 to the third power times 3 to the fourth power?

$$3^3 \cdot 3^4 = ???$$

We reduce this to something we can understand:

$$(3)(3)(3) \cdot (3)(3)(3)(3) = ???$$

After conducting the arithmetic, it looks like this:

$$27 \cdot 81 = 2,187$$

The first thing we notice is that even Little Weenie Numbers can blossom into frighteningly large numbers very quickly when armed with exponents. Larger numbers will exhaust our calculator fingers if we don't come up with some shortcuts. Imagine dealing with things like the speed of light, raised to the 10th power.

The other thing you probably noticed right away is that 2,187 is the same thing as 3 raised to the seventh power.

$$(3)(3)(3)(3)(3)(3)(3) = 2,187$$

In this particular case, rather than conducting all those calculations, we could have said:

$$3^3 \cdot 3^4 = 3^7$$

Although "three to the seventh power" might not be acceptable as the answer to "how many beans are in the jar," in some situations it's all we need. We might be able to manipulate the equation one way or another and never have to perform the arithmetic. Saying "three to the seventh power" is exactly as precise as saying "2,187," and may be much handier.

After doing many problems like this, mathematicians learned that numbers with exponents always work like this. If you're multiplying numbers with the same base, you can simply add up their exponents to get the answer.

$$x^4 \cdot x^{23} = x^{27}$$

For some reason, they use "n" as a common substitute for an unknown exponent when trying to explain this to you. If you see:

$$chimps^n$$

this means "chimps to the nth power" or chimps multiplied times itself an unknown number of times. That funny word, "nth" is pronounced "Enth" and has entered our everyday language, even though we don't all understand what it means. It refers to an unknown exponent.

When books discuss shortcuts for manipulating exponents, and they need to talk about a second exponent, they typically refer to the other exponent as "m." They want to make sure you understand that the two exponents might be different numbers. You could be multiplying "chimps squared" times "chimps cubed," for example. Or times "chimps to the 10th power." They chose "m" and "n" because they look similar enough to cause confusion. Mathematicians consider this good natured fun.

$$chimps^m \cdot chimps^n = chimps^{(m+n)}$$

Reversibility told us that we can reverse a multiplication problem by employing division. We ought to be able to divide chimps$^{(m+n)}$ by chimps^n and wind up with chimps^m. We substitute Little Weenie Numbers and try it.

$$\frac{3^{2+3}}{3^2} = 3^3$$

This is the same as

$$\frac{3^5}{3^2} = 3^3$$

Which is the same as

$$\frac{(3)(3)(3)(3)(3)}{(3)(3)} = (3)(3)(3)$$

If we perform the arithmetic, we get

$$\frac{243}{9} = 27$$

Two-hundred forty-three, divided by 9, is 27. Because 27 = 27, we have evidence that our guess about exponents is correct. Reversibility applies. That's good news. Little Weenie Numbers seem to confirm it. Your teacher will declare it to be true. Absorb it into your personal philosophy. When you multiply numbers with exponents, (and the same base, like chimps, or x, or 15) you just add the exponents.

At first you'll be confused when you multiply. What you'll forget over and over again is that the unstated exponent of "x" is 1. So $(x)(x^2) = x^3$. You have to remember to add that unstated "1." And you won't remember to. I warned you.

If we add exponents when we multiply, does reversibility suggest that we subtract exponents when we divide?

If it did, then apes^5 divided by apes^2 would result in apes^3. Let's try it with Little Weenie numbers:

$$\frac{4^5}{4^2} = 4^{5-2} = 4^3$$

We convert this to arithmetic:

$$\frac{(4)(4)(4)(4)(4)}{(4)(4)} = (4)(4)(4)$$

This is the same as

$$\frac{1024}{16} = 64$$

Our calculators handle the division. One-thousand twenty-four divided by 16 does, in fact, equal 64. Our suspicions have been confirmed. This is the common short-cut method of dividing numbers with exponents: When you divide a base having an exponent by the same base which also has an exponent, you subtract the second exponent from the first one.

But what happens when you divide one number by another with a larger exponent? If you divide baboons^2 by baboons^4 for example? When you subtract 4 from 2, won't you get a negative 2? Can you have a negative exponent? If so, what does baboons$^{(-2)}$ mean? If we assume for a moment that "baboons" stands for "2," then our problem looks like this:

$$\frac{2^2}{2^4}$$

Once again, we transform this into arithmetic:

$$\frac{(2)(2)}{(2)(2)(2)(2)}$$

When we do the multiplication, we get this:

$$\frac{4}{16}$$

Which is the same as

$$\frac{1}{4}$$

This gets us to a simple but odd idea. When the numerator has a smaller exponent than the denominator, your answer will be a fraction. And, because subtracting exponents seems to work so neatly, a fraction will be represented by a negative exponent.

$$\frac{2^2}{2^4} = 2^{-2} = \frac{1}{4}$$

In order for all this to work, and be consistent, and for reversibility to apply, and all the math teachers to be happy, there's one problem that we need to consider. What if both numerator and denominator have the same exponent? What does this little fraction equal?

$$\frac{baboon^3}{baboon^3}$$

You did it in your head, of course. It equals 1. Anything divided by itself equals 1. But what if you apply our short-cut? If we subtract exponents? Subtracting 3 from 3, we get an exponent of zero. Because we know the fraction equals 1, we are compelled, albeit reluctantly, to agree with the odd concept that baboon to the zero power equals 1:

$$\frac{baboon^3}{baboon^3} = baboon^0 = 1$$

And no matter what other number or symbol we try, we get the same result. At some point we concede to the logic of the game. Any number (except zero itself), raised to the zero power equals 1.

If you have a multiplication problem or division problem grouped together, and you raise the whole problem to some power, that's exactly the same as raising each factor of the multiplication or division problem to that power:

$$(ab)^2 = a^2 \bullet b^2$$

and

$$\left(\frac{a}{b}\right)^2 = \left(\frac{a^2}{b^2}\right)$$

If you have a number with an exponent alone or in a group, and you raise that to yet another power, we can use a short-cut. Just multiply the exponents:

$$\left(melon^2\right)^3 = melon^6$$

This may be an opportunity to test the Little Weenie Number process yourself. Substitute some little number for "melon" and try squaring it, then cubing the result. Then try simply multiplying six of your little numbers times each other on your calculator. If you can stand to repeat the experiment with several numbers, you will own the concept and save yourself some brain damage when they cover it in class.

Combining Like Terms

When we sort our pets in the living room, we put our lizards into one box, turtles in another, and tarantulas in another.

But, of course, we also have pets in our bedroom, and (let's face it) in most of the rooms of the house. When we finally assemble all our boxes of critters in one room, we've got several boxes of turtles and of each other pet.

A mathematician would say we've got "a big mess."

It's easier to make sense of our collection if we combine things that are like each other. We'll put all the turtles in one big box, all the lizards in another. Now we only have to count the occupants of each large box. That's easier than counting several boxes, then adding them all up.

This works because we're adding, rather than multiplying. If our collection consisted of lawyers, rather than reptiles, and each one had a different fee schedule, we could not simply throw them all in a box, multiply all their hours times one fee, and have any better understanding than we started with. Of course, it might be fun to try.

Just kidding. We're a lot better off with reptiles.

Things we add or subtract are called terms. Over the course of your fun math activities you're going to wind up with expressions that have several boxes of turtles. That is, you'll have the same species of "terms" scattered across the page. You can make your life a lot easier if you consider putting all your "like terms" into one box.

Unfortunately, it's not as easy to identify "like terms" as it is to identify "turtles." But the principle is the same.

You already know the first category of "like terms." They're regular old garden variety numbers, like "23" and "64." If an expression contains more than one regular old number, and you're only adding or subtracting them, you can throw them into the same box. This is called "combining like terms."

If you see:

$$10 + x + 3$$

you can combine the 10 and the 3 without changing anything. Now it looks like this:

$$13 + x$$

Or, if you see

$$x - 50 + 2x^2 + 50$$

you can combine the (-50) and the (+50) before proceeding further. Once you do, the expression will look like this:

$$x + 2x^2$$

We can also combine any terms in an expression that contain the same unknown.

$$x + 3x + 15 + 2x$$

can be condensed down to

$$6x + 15$$

If x = turtles, we haven't changed our collection at all. We just put them all into the same box.

$$6y + 3y + x - \sqrt{683}$$

means the same thing as

$$9y + x - \sqrt{683}$$

Sometimes an expression will provide several opportunities for such playfulness:

$$2x - y + 14 + x + 2y - 7$$

We can combine our "x" terms, our "y" terms and our regular old numbers. Without changing our little menagerie, we can contain them in fewer cages this way:

$$3x + y + 7$$

Notice this, however, if some of your unknowns include exponents: x^2 is not the same variety of creature as "x." You can't just add them up. Try it with Little Weenie Numbers to convince yourself. If $x = 3$, then $2x^2 + x$ is not at all the same as $3x^2$. Or as $3x$.

On the other hand, $2x^2$ and $3x^2$ are like terms. We combine them and the result is $5x^2$.

Combining like terms is easy if you remember two important rules:

1.   Don't try to combine terms with different exponents;

2.   Don't let any of your lawyers out of the box.

Story Problems

In order to use algebra, the real-world problem must be translated from words into numbers and symbols.

The art of translating, like the game itself, requires practice. The traditional way to provide you this practice is to give you fun little stories called "story problems" or "word problems" and ask you to convert them to equivalent equations.

Folks who write math books live very different lives from you and I, however. They seem to spend a lot of time on trains, for example, which leave cities you and I rarely visit, in hopes of meeting their buddies on other trains at destinations in-between. Sometimes they calculate how much time they can spend on homework and still watch the late news. They launch rockets across rivers, build bridges, and agonize over how tall various trees are. After a couple years of math classes, you'll be uncomfortable hiking through the woods without your calculator handy.

Sometimes they fail to remind you that the process of converting a story into an equation is completely separate from manipulating the equation itself with algebra. Some people become skilled at algebra but panic when presented with a word problem. Others have a knack for converting the word problem, but have no idea what to do with the equation they create.

The first step in subduing a word problem is determining what should be the unknown. Often there are several possibilities.

Perhaps you want to throw a party. You'll need to purchase something for your guests to drink. You decide to buy a keg or two of something, perhaps lemonade. Each keg holds a certain amount of beverage, and costs a certain amount. How much will each guest drink? How many kegs should you buy? How many glasses per keg? How much can you afford? Depending on which of these variables you already know, your unknown might be any of the others.

Some elements of this situation can be expressed in terms of one of the other elements. If Paul always drinks twice as much as Scott, then "Paul equals 2 times Scott." Or, "Scott equals one-half Paul," which you will, of course, abbreviate. There are often many correct ways to translate any problem. In this case, you haven't pinned down any exact numbers, but you have expressed the relationship between the two.

Once we begin manipulating equations, we prefer fewer unknowns. So, the more elements you can express in terms of a single unknown, the happier you'll be later. Perhaps Joey drinks twice as much as Paul, who drinks twice as much as Scott. It would be "correct" to label Scott as "x", Paul as "y" and Joey as "z." Correct, but cumbersome. The total amount they drink might be:

$$x + y + z = total$$

On the other hand, you could decide to call Scott "x" and describe the others in terms of their relationship to him. Paul would be $2x$. Joey would be $4x$. Now we can describe the same problem this way:

$$x + 2x + 4x = total$$

We can conduct several procedures on this equation just the way it sits. We can combine like terms. When we do, we see that

$$7x = total$$

If we can only watch Scott carefully all night, we'll know how much lemonade is being consumed by all three.

When stumped by a word problem, one trick is to employ Little Weenie Numbers. Make some guesses with manageable numbers and see how they play out. Assume the train is moving at 50 miles per hour and see if you wind up in Omaha. Then try 80 miles per hour. After a few embarrassing failures, you may notice a pattern developing or a flaw in your thought process that will lead you to a useful equation.

Another trick is to create a list of possibilities. If the tree is 100 feet tall, the spotted owl squawks every hour. If the tree is 50 feet tall, the owl squawks 10 times an hour. Is this the beginning of a pattern? Or an isolated lunatic bird? If each person drinks 2 lemonades per hour, how many kegs will you need in six hours? Make a little chart showing 1 guest, then 2, then 4 or 10 or whatever number you like. After you've listed several possibilities you may see a relationship which can be expressed as an equation. At the very least you'll sure look busy, scribbling away like a madman, which teachers really like.

These two tricks can't be emphasized too much. Use Little Weenie Numbers and make charts.

Expressions

Mathematicians learn the meaning of many words the same way you learned the meaning of the word "dog." It was never on a vocabulary test, and no teacher explained it. But, after you've been barked at, licked, and chased enough times, pretty soon you absorb the concept.

"Expression" is one of those words. Your teacher honestly believes you already know what it means. She'll casually use it in conversation, and you will smile and nod, as if she were speaking Persian. You'll go home confused, with a mild sense of panic. She'll go home happy, because you seem to be enjoying the class.

When you translate a real-world situation into math, you are "expressing" it in numbers and symbols. We optimistically assume that every bundle of numbers and symbols expresses something. An equal sign or inequality sign shows the relationship between expressions.

All that stuff on the left side of the equal sign is an "expression." All that stuff on the right side is another expression. An expression might include a couple of polynomials, plus a fraction or two, with an exponent thrown in for color. Or, it might be very simple.

Expression or Equation?
Dr. Jim Goes Into More Detail

An expression is some combination of numbers, variables, and operations (like +, -, •, or ÷). It may contain exponents or other operations (like taking a square root), but it won't have symbols like $=$, $\leq$, $\geq$, $<$, or $>$. Nothing that represents the relationship between two expressions. If you know the value of the variables in the "expression" you can replace those letters by the numbers they stand for, and calculate its value.

"Area" can be expressed as the length times the width:

$$L \bullet W$$

If a rug is 6 feet by 4 feet, its area is 24 square feet. We use the expression to translate this into symbols. The expression has no equal sign, but it can have a numerical value.

An equation always has an "equal sign" in it, between the two expressions. Whoever wrote the equation claims that the expression on the left side is equal to the expression on the right side. Equations can only be true or false.

5 = 2 + 3 is a true equation.
1+ 4 = 3 + 3 is a false equation.
Depending on the value of x, equations like

$$x + 3 = 4 - x$$

can be either true or false. It will often be your mission, should you choose to accept it, to find the value or number that makes the equation true.

An inequality always has a symbol like $\leq$, $\geq$, $<$, or $>$. Inequalities are also either true or false and can be "solved."

There is a third type of beast in this part of the forest. We use equal signs to separate different versions of one expression. When we "evaluate an expression," we complete arithmetic, but don't use any of our "equa-tion" tricks, like doing the same thing to both sides.

$$A = L \bullet W = 4 \bullet 6 = 24$$

This is really short-hand for a whole sequence of equations. They would look like this:
$$A = L \bullet W$$
$$L \bullet W = 4 \bullet 6$$
$$4 \bullet 6 = 24$$
We are not solving an equation here. It is hard to locate the right and left sides.

"2 + 3" is an expression and isn't true or false. Expressions aren't true or false. They have some numeri-cal value. "4 + 5 = 8" is an equation and it doesn't have a value. Equations are true or false. They don't have a numerical value.

If you understand these ideas then these two state-ments will seem absurd:

Is it true that 2 + 3?
Find the value of 4 + 5 = 8.

In the first one, we see an expression and we asked if it is true. Expressions aren't true or false. They have some numerical value.

In the second statement, we see an equation and are asked to find its value. Equations are true or false. They don't have a numerical value.

Rational Numbers and Reciprocals

When people first hear the phrase "rational numbers," they picture logical, intelligent numbers, as opposed to the wildly emotional numbers that torment them in their dreams.

No. The word rational comes from a Latin word that means "ratio" or "relationship." Rational people are logical because they see the correct relationships between things. They see the ratios.

Rational numbers express a relationship, like speed compared to distance, or percentage of eaten pizza compared to original pizza. In algebra, relationships and ratios are expressed as fractions.

It's as simple as that. Rational numbers are fractions. Fractions are rational numbers. There is no practical distinction between the two. They look the same, they walk the same. Poke 'em with a stick and they'll squawk the same.

There are two ways to look at a pizza that's been divided. We can compare the number of slices to the original, or we can compare the original to the number of slices. These two different viewpoints are mirror images, or reciprocals of each other. Fifty miles per hour means

$$\frac{50 \; miles}{1 \; hour}$$

Its reciprocal would be $\frac{1 \; hour}{50 \; miles}$ or $\frac{1}{50}$ th of an hour per mile.

Two is the same as $\frac{2}{1}$. It's just a different way to write the same number. So, the reciprocal of two is one-half. This little rule applies to any integer.

The reciprocal of this	is this
32	$\dfrac{1}{32}$
xyz	$\dfrac{1}{xyz}$
$\dfrac{2}{3}$	$\dfrac{3}{2}$
$-76y$	$\dfrac{1}{-76y}$
x^2	$\dfrac{1}{x^2}$

The idea of reciprocals becomes more fun when you're talking about numbers with negative exponents.

Of course, it's absurd to consider a negative exponent, if you're trying to relate this to real life. It is not possible to multiply something times itself a negative number of times. But, to be consistent as a game, we have to deal with the concept. Everything works smoothly if negative exponents lead to the reciprocal of whatever a positive exponent would lead to. So, that's what we agree upon.

If 2^2 equals 4, then 2^{-2} equals one-fourth.

Exponents are just a short-hand way of expressing bulkier numbers. Ten to the fourth power means the same thing as ten-thousand, but you don't use as many zero's to write it. And the reciprocal of that number can be expressed by changing the exponent to a negative number. Ten to the negative fourth power is one-ten-thousandth.

111

Members

The word "member" is a good example of a word that mathematicians use so often they forget that you and I have no idea what they're talking about. If you're talking about "members of an equation" it's simple:

All the stuff to the left of the equal sign is one member. All the stuff to the right is another member. So, when you see:

$$x^2 + 2x = 45 + 2x$$

one member is " $x^2 + 2x$ " and the other member is "45 + 2x" Every equation has two members.

This becomes a tad more confusing because they also use the word "member" to refer to one of the items within a "set." A set is a defined group. Tenors is a group of guys with high voices and bad attitudes. "Tenors" is the set, each guy with a high voice and a bad attitude is a "member" of the set.

Usually you'll know if someone's talking about a "member of an equation" or a "member of a set." The easy way to tell the difference is to notice the name of the chapter you're studying. If you're studying sets, you're probably talking about members of a set. Otherwise, you're probably talking about members of an equation.

Adding and Subtracting as Strategy

Every Saturday I like to weigh my spider collection. I put the jar containing them on one side of the scale, put an empty jar on the other side, and add little weights until it's perfectly balanced. Once balanced, I know my spiders are as heavy as the little weights on the other side. Because the weights are each clearly labeled, I know how much my tiny prizes weigh. By now, of course, I hunger to express this as an equation:

$$Spiders + Jar = Weights + Jar$$

The reason this works is that both jars weigh the same amount. They don't affect the balance of the scale. Of course, they don't tell me much about the weight of the spiders, either.

I can add identical weights to each side of the scale without tipping it. I can also subtract equal weights. I can remove a jar from each side without tipping the scale. So I can subtract the weight of each jar from both sides to see how much the spiders themselves weigh.

$$
\begin{array}{ll}
Spiders + Jar & = Weights + Jar \\
\qquad\quad - Jar & \qquad\qquad - Jar \\
\hline
Spiders & = Weights
\end{array}
$$

Last Saturday, I was dismayed to see that someone (probably my sister) had dumped 14 paper clips into my spider jar. Now, I like spiders as much as the next guy, but I wasn't crazy about sticking my hand in there to pull out the paper clips. On the other hand, it wouldn't be Saturday if I couldn't weigh my pets.

It occurred to me that I could simply add 14 paper clips to the empty jar before I added any weights. If there were 14 paper clips on each side of the scale, the thing should balance, and any difference would be my spiders. That is, we took the original equation:

$$Spiders + Jar = Weights + Jar$$

Since my sister added 14 paper clips to the left side, I could keep everything balanced by adding 14 paper clips to the right side:

$$Spiders + Jar + 14\ paper\ clips = Jar + Weights + 14\ Paper\ clips$$

So, how much do my little beauties weigh? To find out, I subtract the weight of the paper clips from each side:

$$
\begin{array}{ll}
Spiders + Jar + 14\ Paper\ clips = Jar + Weights + 14\ Paper\ clips \\
\quad\quad\quad\quad\quad - 14\ Paper\ clips \quad\quad\quad\quad\quad\quad - 14\ Paper\ clips \\
\hline
Spiders + Jar \quad\quad\quad\quad = Jar + Weights
\end{array}
$$

Once again, I subtract *Jar* from both side. The spiders weigh exactly the same amount as the weights. Because the weights are labeled, my Saturday is complete.

You don't change the truth of an equation if you add or subtract the same thing to each side. Perhaps some day you'll see an equation that looks like this:

$$(10 + z) - 3 = x\text{-}3$$

Add 3 to both sides of this equation, and we get:

$$10 + z = x$$

Maybe we want to know what z is. If we subtract 10 from both sides of the equation, we realize that $z = x - 10$. If it's confusing, try substituting some Little Weenie Numbers, and see what happens.

Changing Signs

If you want to eliminate 35 from one side of an equation, you simply subtract 35 from both sides:

$$
\begin{array}{rl}
x + 35 = & 50 \\
-35 = & -35 \\
\hline
x \quad = & 15
\end{array}
$$

This idea works for any number. If you wanted to eliminate a negative 10 from one side, you add positive 10 to both sides.

$$
\begin{array}{ll}
xyz \; -10 & = AB \\
xyz - 10 + 10 & = AB + 10 \\
\hline
xyz & = AB + 10
\end{array}
$$

This little maneuver is done so frequently, after a while it becomes second nature. In fact, people forget that's what they're really doing. Instead, they say that you can move a number from one side of an equation to the other, and when you do, the sign changes. Positive numbers become negative numbers when moved across the equal sign, and vice versa. It is this move that the ancient Arabic word "al-jabr" describes. That's where the word "algebra" comes from. At least, according to some books.

In our first example, they'd say they moved the 35 to the other side of the equation, and when they did, they had to change the sign.

Thinking about it this way works just as well. But if you memorize it as a rule, and apply it without thought, it's only one more magical incantation. There is a limit to the number of rules a person can learn and follow, if none of the rules make much sense. There are no mystical spells in algebra. Changing signs is an extension of our basic strategy: If we do the same thing to both sides of an equation, the scale won't tip, the equality will be preserved. Once you're comfortable with that concept, the short-cut of changing signs won't hurt you.

The neatest application of changing signs is if both sides of an equation are negative numbers. For example,

$$(-AB) = (-45)$$

If we add AB to each side of the equation, we get:

$$(-AB) + AB = (-45) + AB$$

which results in

$$0 = (-45) + AB$$

Then we add 45 to each side:

$$0 + 45 = 45 - 45 + AB$$

$$45 = AB$$

After you've done about three thousand problems just like this one, you realize there's a certain efficiency to skipping some steps on the way to your inevitable conclusion. You'll move each number across the equal sign, the negatives will become positives, and the positives become negatives.

Another approach is to multiply both sides by a negative 1:

$$(-1)(-AB) = (-1)(-45)$$

Which results in:

$$AB = 45$$

It's a different technique, but the result is the same. You choose the method you like.

Two Variables

Adding or subtracting the same thing to each side of an equation doesn't change its equality.

Less obvious is that we can add two identical quantities to each side, even if we call them by different names.

If Tom and John are identical twins and weigh exactly the same amount, we can add Tom to one side of our scale, and John to the other, and the scale won't tip.

If x equals 14, we could add x to one side of the equation, and 14 to the other.

Sometimes a problem contains two unknowns. We may know enough to create two equations but, because each equation contains both unknowns, we can't solve either one alone. For example,

"I'm thinking of two numbers that add up to 13. Their difference is 3."

Your first question ought to be "Does that poor guy have a life?" Later, you notice that the first part leads to the equation

$$x + y = 13$$

And the second part can be written

$$x - y = 3$$

The problem $x + y = 13$ can't be solved by itself. Neither can the problem $x - y = 3$. But, if both are true, we have a shot.

Because $x - y$ is the same thing as 3, as our problem says, we add $(x - y)$ to the left member, and 3 to the right member of our first equation:

$$
\begin{array}{rcr}
x + y & = & 13 \\
+\,x - y & & +\,3 \\
\hline
2x + 0 & = & 16
\end{array}
$$

Once we know that $2x$ equals 16, we can divide both sides by 2 and learn that x equals 8. Once we know what x is, we can figure out y pretty easily.

Dealing with two variables isn't uncommon. Most of us are confused about more than one thing at a time.

For example: Your girlfriend calls and states that she spent the last 2 hours studying algebra and talking on the phone to her ex-boyfriend. Right away, you're confused about several things. What did she ever see in him, anyway? Why is she talking to him? Didn't she say he moved to China? Does she understand algebra well enough to help you? Does her ex-boyfriend?

But, when you ask casual questions, trying to understand these weighty matters, she reassures you. She says she spent 30 minutes more studying than she did on the phone.

Your first instinct, of course, is to write an equation. But there are two unknowns: How long did she spend studying, and how long did she flirt with that bum? You might assign letters to each of these:

$$x = \text{time on phone}$$
$$y = \text{time studying}$$

We know that $x + y = 120$ minutes
And that $y = x + 30$

We'd like all our unknowns on one side of that second equation. We subtract x from both sides of "$y = x + 30$" and get:

$$y - x = 30$$

We take our first equation:

$$x + y = 120 \text{ minutes}$$

We want to add the same thing to both sides. We refer back to our second equation. Because y - x equals thirty, we can add y - x to one side of the first equation, and 30 to the other side, and the scale of equality will not waver.

$$
\begin{array}{rcl}
y + x & = & 120 \text{ minutes} \\
+ \quad y - x & = & 30 \text{ minutes} \\
\hline
2y & = & 150 \text{ minutes}
\end{array}
$$

Dividing both sides by 2 gets us "y equals seventy-five." She studied for seventy-five minutes. She only spent forty-five minutes talking to the jerk. You feel a lot better.

Rearranging

Sometimes a choir director may find it useful to put all the tenors together in one section. It's easier to keep an eye on them. Other times, to prevent them from giggling at each other, she may decide to break them up and intersperse them in the other sections. The music produced will not be affected. They'll still sing the tenor part, and the sound of the choir won't be affected. Each arrangement aids a particular strategy.

In algebra, we reorganize problems to make them convenient for specific strategies. There are three principles to keep in mind when attempting this. You will learn these as the commutative, associative, and distributive principles. Giving them fancy names keeps the tenors from understanding that they're being moved.

First, it doesn't matter what order you add things if that's all you're doing. But some arrangements might be more convenient for you.

The same is true of multiplication. If you've got a whole truckful of numbers to multiply times each other, and that's all you're going to do with them, you don't have to worry about which one you pull off the truck first. The answer will always be the same.

Another way to think of this is that it doesn't matter which direction you perform addition or multiplication. If you're adding a long column of numbers, you can add from top to bottom or bottom to top. If you've got a string of numbers to multiply written horizontally, like an equation, it doesn't matter if you work from right to left, or left to right, or if you start in the middle somewhere. Mathematicians call this the commutative principle. You can remember the name because it's like commuting on a train or bus. The fare's the same, regardless of which direction you're heading.

The commutative principle works for addition and multiplication, but not subtraction or division. Obviously, you can't switch the order of a subtraction problem, or of a division problem. Ten minus 3 is different from 3 minus 10.

You can reorganize a problem if the commutative principle indicates that your new version equals the original. For example:

$$(10)(100)(3)(4) = ???$$

might be easier if you work from right to left. This multiplication problem:

$$(3)(10)(4)(100)$$

might be easier if you rearrange it to look like this

$$(3)(4)(10)(100)$$

Doing so won't affect the result.

The commutative principle works within any

group of numbers you're adding or multiplying. But only within the group. Once you begin a different arithmetic operation, you've begun a new group of numbers. So, if you see this:

$$(100)(10)(3)(4) + (100)(10)(3)(4) = ???$$

you can rearrange within the group to the left of the plus sign, or within the group to the right of the plus sign. But you couldn't put something from one group into the other. In this example:

$$(433 + 5 + 2)(433 + 5 + 2) = ???$$

you can rearrange the terms within either parentheses, but you can't mix in something from the other group. Besides grouping the terms together, the parentheses indicates that you're multiplying one group times the other. The change from adding to multiplying warns us of the end of a group we can rearrange.

But if we were adding the two groups together, like this
$$(433 + 5 + 2) + (433 + 5 + 2) = ???$$

we really just have one big happy addition problem that we can rearrange until our pigs come home, if we want. There is no change of arithmetic operation. It's all addition.

So, one more time, the commutative principle says: "The order you add numbers is irrelevant, as long as that's all you're doing. Same thing for multiplication."

In that last example, where the entire problem was addition, the parentheses were also irrelevant. They could have been anywhere. Parentheses tell us which things are grouped. In this case, it didn't matter. We can add in any order we desire. The answer won't change. All these are the same:

$$(433 + 5 + 2) + (433 + 5 + 2) = ???$$
$$(433 + 5 + 2 + 433 + 5 + 2) = ???$$
$$433 + (5 + 2 + 433) + (5 + 2) = ???$$
$$433 + 5 + (2 + 433 + 5 + 2) = ???$$
$$(433 + 5 + 2) + (433 + 5) + 2 = ???$$

The same thing applies to multiplication. We can group factors together however we want, as long as we're only multiplying. These are all the same:

$$4332 \cdot 123x \cdot 19 \cdot 22 = ???$$
$$(4332 \cdot 123x)19 \cdot 22 = ???$$
$$4332 (123x \cdot 19)22 = ???$$
$$4332 (123x \cdot 19 \cdot 22) = ???$$
$$(4332 \cdot 123x)(19 \cdot 22) = ???$$

In fact, "123x" means "123 times x." So, we could have grouped the x with the 19, instead of the 123.

You can think of this grouping process as forcing numbers to associate with each other. A group is an association. And the idea that grouping doesn't matter within an addition or multiplication problem is called the associative principle.

Sometimes things enter a problem grouped a certain way: Tom is 1 year older than Mike. So Tom's age is always M + 1. To keep that idea clear, Tom's age might be kept within neat parentheses, like this:

$$T = (M + 1)$$

Later, when you're performing your cruel tricks

on the entire equation, you might discover some strategic advantage to moving, or removing, the parentheses the package came wrapped in. As long as you keep the associative principle in mind, this won't affect your result. For example, Tom's age might show up in this expression:

$$(23 + M) + (M + 1) + 4$$

You may decide it makes more sense, for some diabolical reason, to group the M s together. Because all we're doing in the problem is adding, this is legal:

$$23 + (M + M) + 1 + 4$$

The commutative principle and the associative principle become much more powerful and useful when used together. Remembering their limits, we can rearrange and regroup problems most handily. If we see:

$$53x + 16y + 12x + 33 + x + y + 16$$

our first healthy reaction is panic. But, since the entire bucketful involves only addition, we know we can put them in any order, and group them however we choose. So we might transform it into something like this:

$$(53x + 12x + x) + (16y + y) + (33 + 16)$$

Sure, we still have no idea what to do with them, but somehow that doesn't bother us as much now.

Both the commutative principle and the associative principle apply to addition and multiplication.

The distributive principle allows us to rearrange some problems involving both.

If you decide to sell two investments to pay for a new boat, you add the money together, and let the boat dealer relieve you of your cash. If each of your investments doubles in value before you liquidate, you take twice as much money down to the marina, although, in the real world, you don't get change. You get a bigger boat. Let's use my own portfolio to illustrate:

Double(Piggy Bank One + Piggy Bank Two) = Bigger Boat

It doesn't matter if you combine your accounts first, and that one larger account doubles, or if each account doubles, then you combine them. Same amount of money.

(Double Piggy Bank One) + (Double Piggy Bank Two) =
Same Bigger Boat

This is not some difficult mathematical abstraction. It's common sense. Still, once the numbers get larger, and everything looks more complex, it's easy to forget that you have known this for years. We can take the same information and make it look more confusing (which, I am convinced, is one of the secret goals of mathematicians). We'll say the amount of money available to us is x, rather than "boat." We'll call our two investments "A" and "B." Doubling our money is the same as multiplying it by 2. Now we can restate our simple little boat problem and make it look like this:

$$x = 2(A + B)$$

Now we step away from the equation as if it might

be rabid. What can we do with this? Nothing. It's an equation, for heaven's sake. And we're just beginners. Put it back in its cage.

But, of course, it's the same thing we just discussed.

$$x = 2(A + B)$$

is the same as

$$x = (2A) + (2B)$$

Your Real Algebra Book won't use "2" when it explains this, but will substitute another generic letter, to indicate that any number will work. Now it's a "formula." So, you'll probably see the distributive principle described something like this:

$$a(b + c) = ab + ac$$

This concept will be so obvious to your teacher, she will probably spend no time at all on it. You'll memorize "distributive property," and this formula, which is a lot like memorizing the names of fishing holes, but never getting to fish.

The way to understand and believe this (or any other) concept is to substitute Little Weenie Numbers for a, b, and c, and see what kind of fish bite for you.

The distributive principle is a handy tool for reorganizing. As you gaze at the last example, you may notice one interesting transformation that occurred when we used it.

$$a(b+c)$$

is a multiplication problem, while

$$ab + ac$$

is an addition problem. Using the distributive principle we can translate a multiplication problem into an addition problem, and vice versa. Changing addition problems into multiplication problems is called "factoring." If we see something like this:

$$a(b + c) + ab + ac = ???$$

we may wish we could rearrange it. But the commutative principle and associative principle are helpless, since we're not only adding, or only multiplying. We're doing both. The distributive, however, can transform that first group from a multiplication problem to an addition problem. Now the whole expression looks like this:

$$ab + ac + ab + ac = ????$$

If we want, we can rearrange it further because of the commutative property:

$$ab + ab + ac + ac$$

And, if we want to group with the associative property, we are free to do so:

$$(ab + ab) + (ac + ac)$$

We could even complete some arithmetic, if we wanted, and come up with this:

$$2ab + 2ac$$

which is identical to what we started with.

Each of these principles are tools. Use them as you choose in a problem, or decline to use them. Just because you can legally rearrange or regroup doesn't mean that's the strategy that will solve a problem. There is not necessarily only one correct approach.

Cross Multiplying

If two fractions are equal, you can multiply each numerator times the denominator of the other, and the two results will equal each other. That's cross multiplying.

"Cross multiplying" is a powerful little shortcut that solves many problems like magic. It's so easy, you'll learn it in thirty seconds. What will take longer is remembering when to use it. And those of us who like to understand why things work can easily lose half a semester trying to make sense of it.

Cross multiplying is a way of manipulating two equal fractions. That's what people forget. They try to use it to multiply fractions, or to find common denominators, or to heal insect bites. All wrong. Cross multiplying is a way of manipulating two equal fractions. The fractions themselves do not survive this operation. The only thing that survives is the equality. When all the smoke clears, you no longer have a fraction at all. In fact, cross-multiplying can be a useful tool to get rid of fractions. When you're done, there are two completely different characters on each side of the equal sign. These new characters appear to have little to do with the original fractions. But they still equal each other.

$$\frac{8}{12} = \frac{2}{3}$$

This example contains no mysteries. It's a simple statement of fact, which we assume to be truthful. It employs Little Weenie Numbers as laboratory rats, before applying the procedure to full-grown and possibly dangerous polynomials.

We multiply each of these fractions by some variety of "1." To ensure that each final denominator will be a multiple of the other, we construct our fraction version of "1" from the denominator of the other fraction. That is, we multiply 8/12 times 3/3 and we multiply 2/3 by 12/12.

$$\frac{8}{12} \bullet \frac{3}{3} = \frac{24}{36}$$

and

$$\frac{2}{3} \bullet \frac{12}{12} = \frac{24}{36}$$

Two important events happen during this operation. First, we get a common denominator. This makes it possible for us to consider and compare the meaningful aspects of the fractions.

Second, once we've got that common denominator, we ignore it. The pizza has been sliced a certain way. Now our concern becomes what we do with the slices. Twenty-four equals twenty-four, and that's the significant, useful part of this little game.

In fact, if we wanted, we could multiply both sides of our "equation" by thirty-six and eliminate the denominators altogether. If we did, we'd see this:

$$\frac{24}{36} \bullet 36 = \frac{24}{36} \bullet 36$$

which leads us to:

$$24 = 24$$

Our original fractions did not survive these manipulations. "24" is not the same thing as "24 over thirty-six." But the scale never tipped. 24 does equal 24, just as our original fractions were equal to each other.

After doing thousands of these pups, mathematicians realized they could eliminate a bunch of steps. They threw away the results of most of the preliminary steps anyway. Why not cut to the chase? Why not skip all the steps that led to the disposable common denominator, and keep only the steps that lead to the final numerators? Would some procedure maintain the purity of the entire process but eliminate several steps of arithmetic?

Yes. Cross multiplying.

In our example, you'll notice that multiplying each numerator times the other fraction's denominator gives us 24. 2 times 12 equals 24, and 3 times 8 equals 24. That's what we came up with after spending too much of a pleasant afternoon handling the nasty things. And it always works like that.

That's why we can simply "cross multiply" to convert equivalent fractions to a different pair of equal items.

If this is true:

$$\frac{rats}{kangaroos} = \frac{corndog}{lawyer}$$

Then this is also true:

$$rats\ (lawyers) = kangaroos(corndog)$$

Different, but still true.

This all seems so harmless and silly as long as we're working with Little Weenie Numbers. You may find it obvious and boring. Until you're cornered in a dark alley by something that looks like this:

$$\frac{x+45}{yz} = \frac{x}{432}$$

You will try to remember your best moves, and wonder if the beast has a soft, vulnerable spot. Unfortunately, you thought cross multiplying was sissy stuff that you've outgrown, so you skipped through this section too quickly. Now, when it could help you, you've left it in your other jeans.

Or, perhaps, you remember that you can cross multiply equal fractions without affecting the equality of the statement. You try it.

After cross multiplying, you would have a second statement that is equally true. Different, remember, but just as true:

$$(x+45) \bullet 432 = xyz$$

Now you're on your way. You see other strategies that might actually solve this equation. You have eliminated the complication of dealing with fractions.

The only difficult thing about cross multiplying is remembering when to do it. You can do it whenever you have two fractions equal to each other. And, of course, if a tremendous urge to cross multiply comes over you, it is usually possible to express any number as a fraction. x, for example, can always be described as x/1 .

You have to multiply the entire numerator of each fraction by the entire denominator of the other fraction. Often, the easiest way to ensure you do is by placing them within parentheses:

$$\frac{pigs + slop}{sheep - dip} = \frac{mess}{plea + bargain}$$

When you cross multiply something like this, parentheses can help you keep track of what you intend to do:

(Pigs + slop)(plea + bargain) = (sheep - dip)(mess)

Cross multiplying is a neat trick when two fractions equal each other. But that's not all it's good for. Some day you will encounter two unequal fractions and want to play with them. Cross multiplying will once again prove useful. To prepare you for that shining moment, it is useful to get into this habit: When you cross multiply, keep your numerators on their original side of the equation.

Cross Multiplying: Dr. Jim Expands

If I had asked you yesterday to explain why cross multiplying works you probably wouldn't have been able to answer. For me, as a math teacher, this symbolizes how we systematically rob people of the chance to appreciate how cool numbers and fractions are, and how they work together. At some point you learned it, you've used it, but you don't understand it or love it.

By the time you read the next couple of pages you'll see how cross multiplying works and why I suggest that you never use it.

We need to be able to decide when fractions are equal. This is especially important in algebra, where dangerous unknowns may lurk in the shadows. For example, let's consider this "equivalence" :

$$\frac{x}{12} = \frac{2}{3}$$

Using the "squint" method you can probably guess that x should be 8 in order to make the equation true. Some of you may remember that "THE" way to do this problem is to cross multiply; that is, set the product, x times 3, equal to the product, 2 times 12. We would get

$$3x = 24.$$

Dividing both sides by 3 ... sure enough, we get x = 8. It is pretty cool the way it works. The question is: Do you know why that step worked? If you do, then you know that algebra is something that can make sense. If you don't, then learning this method will contribute to your belief that algebra is mysterious and magic.

Lots of people use cross multiplying incorrectly to attempt this one:

$$\frac{x}{12} + \frac{2}{3}$$

and get completely lost when they need to solve this one:

$$\frac{x}{12} + \frac{2}{3} = \frac{x}{5}$$

So let me suggest an alternative to learning how to cross multiply. If you know about it, forget it completely! You'll never need it. If you don't know it, learn the following system. It makes lots of sense and there is nothing to remember. We'll just use the principles

Make the problem simpler!
and
Do to one side of an equation as you would the other.

133

Carve these principles into your favorite stone.

Let's look at this equation:

$$\frac{x}{12} = \frac{2}{3}$$

Your first reaction may be "I don't like fractions!" So the first goal would be to get rid of them. (The only cool thing about fractions is that you can almost always get rid of them.) After you get over your reaction, notice the left fraction . We can get rid of that denominator by multiplying by 12. Whatever we do to one side, we must do to the other. Now it looks like this:

$$12 \bullet \frac{x}{12} = 12 \bullet \frac{2}{3}$$

So,

$$x = 12 \bullet \frac{2}{3} = \frac{12 \bullet 2}{3} = \frac{24}{3} = 8$$

The solution is x = 8. No steps to memorize.

Now let's look at the second equation we mentioned earlier.

$$\frac{x}{12} + \frac{2}{3} = \frac{x}{5}$$

There are several things about it that we don't like. For one thing, it has 12, 3, and 5 in the denominators. We don't like fractions. So the first job is to get rid of them. Let's get rid of the 12 first:

$$12 \bullet \left(\frac{x}{12} + \frac{2}{3} \right) = 12 \bullet \left(\frac{x}{5} \right)$$

Using the distributive principle we get

$$x + \frac{12 \bullet 2}{3} = \frac{12 \bullet x}{5}$$

or

$$x + \frac{24}{3} = \frac{12x}{5}$$

Finally, it is:

$$x + 8 = \frac{12x}{5}$$

OK. That step is done. We got rid of the fraction with a denominator of 12. It took care of the fraction with a denominator of 3 at the same time. Still don't like the fraction with a 5 denominator? No problem. Multiply both sides by 5 and get

$$5 \bullet (x + 8) = 5 \bullet \left(\frac{12x}{5} \right)$$

and so it becomes

$$5x + 40 = 12x$$

Now before we rush off and finish this one, notice that we didn't need any magic. We just paid attention to what we didn't like about the problem, used THE principles to get rid of those aspects. And we got a nice little equation . . . Well, at least it doesn't have any fractions in it.

In solving that last equation, the next thing that we notice is that there are x's on both sides. Our work is nearly done when there are x's only on one side. We subtract 5x from both sides and get

$$40 = 7x.$$

And finally, divide both sides by 7. So, x equals

$$\frac{40}{7}$$

On my calculator, that is about 5.714.

Squares and Square Roots

When you multiply any number by itself, you have squared it. 2 times 2 equals 4. You could also say: "2 squared equals 4."

Multiplying a number by itself happens so frequently, there's a special short-hand to describe it. A little 2, placed above and to the right of a number means "squared," or "multiplied by itself." The 2 is called an exponent.

The arithmetic isn't any more difficult than any other multiplication problem. But it can be intimidating to see all these little "twos" sprinkled across a page, especially if we also have many "x's" and "y's" scattered like tiny land mines among the numbers.

"Ten times itself equals 100." This entire sentence can be abbreviated 10^2. Lampshade times Lampshade could be abbreviated "Lampshade ^{2}" and we would read it "Lampshade Squared." You might describe someone's level of sensitivity in terms of a hockey puck. To indicate a level of sensitivity equal to a hockey puck times itself, you might say "hockey puck squared," and you would write it "(hockey puck)^2." This is an example of how algebra can be useful in your everyday life.

It is possible to multiply an unknown times itself:

$$(x)(x) = ???$$

This can be abbreviated x^2. Whatever x represents, whether it's "4" or "hockey puck" or "1/235", we know that it is multiplied by itself. We don't need to panic just because we don't know what x represents. Squared numbers act like any other number in an equation.

Reversibility has not abandoned us, although it has become a little trickier. The reverse operation of squaring is taking the square root. 2 times 2 equals 4. We could say: two squared is four. If we start at the cookie, instead of the recipe, we say "the square root of 4 is 2." The wonderfully intimidating symbol for "taking the square root of" something looks like this:

It's called a "radical sign."

Because 2 times 2 is 4, the square root of 4 is 2. Since 3 times 3 is 9, the square root of 9 is 3. To determine the square root of something, we begin with square roots that we know and experiment. For the purposes of this book, here's a little chart. I promise not to use any square roots that aren't on the chart:

The square root of:	is this:
1	1
4	2
9	3
16	4
25	5
36	6
49	7
64	8
81	9
100	10
121	11
361	19
cow droppings	pig spit
jabberwocky	slithy toves
x^2	x
y^2	y

Although any number can be an exponent, 2 is by far the most common. You will see squares over and over again, and will learn whole techniques that only apply to equations that involve squares and square roots. You will spend much less time worrying about other exponents.

A note about the chart: Technically, the square root of x^2 is x only if x started out as a positive number. x^2 will be 25 if x equals either 5 or (-5) but they only call positive numbers square roots. It does not seem to bother mathematicians at all to qualify their rules this way, but it may drive you crazy. Similarly, the square root of cow droppings is pig spit, even though (pig spit)^2 and (-pig spit)^2 both equal cow droppings.

Solving Linear and Quadratic Equations

The words "linear" and "quadratic" were originally created to frighten children. Now, mathematicians use them when showing off to friends, trying to impress people of the opposite sex, and in job interviews. Like most other tools used for these purposes, the truth is much less impressive.

They refer to exponents.

If an equation doesn't have any exponents, it's a linear equation. That is, if nothing is squared, or cubed (which means it has an exponent of 3), or raised to some other power. For example:

$$x + 3 = 93$$

is a linear equation. No exponents. Linear equations are also called "first degree equations."

If one of the variables in an equation is squared, it's a quadratic equation, or "second degree equation." For example:

$$x^2 + 3 = 93$$

is a quadratic equation. One of the variables is squared, but nothing is raised to any higher exponent than 2.

That's it.

When you use the "graphic approach" to algebra, and describe equations as lines on a chart, a linear equation with two variables (like x and y) will produce a straight line. A quadratic with two variables (one of them squared) will produce a curved line called a "parabola."

Most algebra problems are either linear or quadratic. Solving "cubic" equations (where one of the terms is raised to the third power, or has an exponent of 3) is more cumbersome. Textbook authors struggle to find neat, solvable cubic equations to throw in. Once you're comfortable with linear and quadratic equations, you're probably going to pass algebra.

Solving a linear equation calls for one set of strategies. Solving a quadratic calls for a different set of strategies.

To solve a linear equation, we apply arithmetic equally to each side of an equation. We add the same amount to both sides, or we divide both sides by the same amount. Perhaps we multiply or subtract. And sometimes we re-arrange the terms for our convenience.

In addition to these basics, there are a couple of shortcuts. We can multiply or divide anything by "1" without changing it, and we can express "1" in a jillion ways. We can cross multiply. That's our complete bag of tricks for dealing with linear equations.

If you have trouble with linear equations, after a reasonable amount of practice, it's probably because:

     A) you've forgotten how fractions behave;
     B) negative numbers confuse you;
     C) unknowns (like "x" or "y") confuse you; or
     D) you aren't sure of the legal ways to rearrange

terms.

In a quadratic equation, something is squared, but nothing is raised to a higher power than that. Some very simple quadratic equations will yield to your linear strategies. Most of the time, you treat them differently.

There are two basic strategies for solving a quadratic equation:
1. Factoring
2. The Quadratic Formula.

"Factoring" means manipulating expressions until they're in the form of a multiplication problem. If you can rearrange an equation so that it's a multiplication problem equal to zero, it's simple to identify the unknown. If a problem can be solved by factoring, that's the slick, fast way to handle it.

But not all quadratics can be solved by factoring. Unfortunately, you may not realize this until you've spent time trying. Some people hate to risk that time and effort. Other people hate to factor, or never really got the hang of it. On the other hand, many people find factoring to be a fun process, a game, and they'd rather factor than eat ice cream. Fortunately for the rest of us, most of these people live, or teach, in institutions.

Your alternative to factoring, the quadratic formula, is a frightening-looking beast. It's an equation with "a's" and "b's" and a square root sign (formally known as a radical sign) all mixed together in a big fraction. You plug the information from the equation you're trying to solve into the quadratic formula, complete the arithmetic, and you've solved your equation. Of course, you've got to learn how to plug information into it. Plus, it's a nasty enough critter that it will take more than thirty seconds to memorize. And, even after you've translated your problem into the quadratic formula, you'll give your calculator a work-out completing the arithmetic.

But it always works. So you have a relatively quick way to solve quadratics (factoring) that doesn't always work and a lengthier process (the quadratic formula) that always does work.

Although you may have to deal with some equations grander than quadratics, you won't have to learn a "Cubic Formula," and you won't have to learn a vast new landscape. Most algebra problems you're likely to encounter soon will be either linear or quadratic. In fact, the orderly, predictable patterns we learn to recognize at this level break down as the exponents get larger. Quadratic equations are not the beginning of a series of increasingly difficult problems. They're the end of the line: The most difficult of the species that you're likely to encounter.

So, take heart from the fact that the quadratic formula always works. It's the most difficult thing you're likely to encounter this year.

Reversibility as a Strategy

You can use the principle of reversibility to shape equations to your taste. If you know how you got somewhere, it's easy to figure out how to get back. More times than you'd guess, if you think about it, you know how to reverse a mathematical process.

For example, consider fractions. If you divide "1" by "4," you get the fraction one-fourth. If you multiply one-fourth by 4, you get 1. That's reversibility. It's obvious with small numbers like this, but it works equally well with large numbers, or unknown numbers. If "1" is the numerator, multiply a fraction by its denominator, the answer is one:

$$\frac{1}{12} \bullet 12 = 1 \qquad\qquad \frac{1}{345} \bullet 345 = 1$$

$$\frac{1}{x} \bullet x = 1$$

$$\frac{1}{fishsandwich} \bullet fishsandwich = 1$$

What happens if the numerator is not 1? Then you get the original numerator, and the denominator simply disappears:

$$\frac{23}{567} \bullet 567 = 23$$

$$\frac{Van\ Gogh}{Picasso} \bullet Picasso = Van\ Gogh$$

Or, as Stephen King would say:

$$\frac{35x + 63z}{19} \bullet 19 = 35x + 63z$$

When you multiply a fraction by any number, including a number that happens to be the denominator, you change the fraction. One-fourth is not the same as 4. So, if you have a fraction on one side of an equation, and you'd rather have no fractions at all, multiply both sides of the equation by the same amount. You'll change both sides equally.

$$\frac{1}{4} = ten \ \ rabbits$$

multiply both sides by four:

$$4 \bullet \frac{1}{4} = 4 \bullet \ \ ten \ \ rabbits$$

The old and new equations do not say the same thing, of course. You've changed both sides of the equation. 4 times 1/4 equals 1, which is completely different from 1/4. And 4 times 10 rabbits is obviously not the same as 10 rabbits.

What you *have* done is preserved the equality between the two sides. The scale still balances. You've transformed one true statement into a different, but equally true, statement.

When you complete the arithmetic, you see that 4 times one-fourth equals 1, and 4 times 10 rabbits equals 40 rabbits. The answer looks like this:

$$1 = 40 \ rabbits.$$

This is not enough information for you to create a mental picture. What is 1? A box of rabbits? A shipment? A dollars worth? Are we saying "1 litter equals forty rabbits?" There's no way of knowing. And it's not important. We translate life into equations before we begin conducting our algebraic manipulations upon it. Once it enters the arena in its math costume, reality becomes an irrelevant distraction.

Suppose we see an equation like this:

$$\frac{pizza + pop}{Winnebago} = catfish$$

This is a very advanced formula, because it involves four variables and makes no sense. You don't usually see formulas this ridiculous until graduate school. But we can already do some things to it. We can get rid of the fraction. Because the denominator is Winnebagos, all we have to do is multiply both sides of the equation by Winnebagos:

$$Winnebagos \bullet \frac{pizza + pop}{Winnebagos} = (catfish)(Winnebagos)$$

Catfish times Winnebagos is certainly different than Catfish all by themselves. Doesn't matter. We have done the same thing to both sides of the equation, the equality is preserved, the scale has not been tipped. Math has taught us that:

$$Pizza + Pop = (Catfish)(Winnebagos)$$

What if there are fractions on both sides of the equation? No problem. Just deal with them one at a time.

145

Another example:

$$\frac{1}{3}(x) = \frac{1}{4}(y)$$

First we multiply both sides by 3 to clean up the left side:

$$3 \bullet \frac{1}{3}(x) = 3 \bullet \frac{1}{4}(y)$$

This is the same as

$$\frac{3}{1} \bullet \frac{1}{3}(x) = \frac{3}{1} \bullet \frac{1}{4}(y)$$

This is the same as

$$1x = \frac{3}{4}(y)$$

Now we multiply both sides by 4 to get rid of the remaining fraction:

$$\frac{4}{1} \bullet \frac{x}{1} = \frac{4}{1} \bullet \frac{3}{4}(y)$$

This gets us to

$$4x = 3y$$

Perhaps this doesn't identify exactly what x and y are. But, if we started from a recipe in a book, and x stands for sugar while y stands for flour, now we see the relationship between sugar and flour in our cookies. If I throw 4 cups of sugar in, I better throw 3 cups of flour in as well. If I use 40 cups of sugar, I have to use 30 cups of flour.

If I see this equation:

$$\frac{red}{blue} = \frac{green}{purple}$$

I can eliminate the fractions, even though, as is usually the case, I have no clue what the problem is talking about:

$$\frac{blue}{1} \bullet \frac{red}{blue} = \frac{blue}{1} \bullet \frac{green}{purple}$$

The blues on the left "cancel each other out." They reverse each other. This leads us to:

$$red = \frac{blue}{1} \bullet \frac{green}{purple}$$

To get rid of the final fraction, we can multiply both sides of the equation by purple. This gives us:

$$(red)(purple) = (blue)(green)$$

Still doesn't make much sense, but we did what we set out to do. We maintained the equality of the equation while we got rid of the fractions.

Squaring & Square Rooting as Strategies

To reverse something that's been squared, we take the square root of it. To reverse a square root, we square it. This is another example of reversibility at work for you. We can use it to our tactical advantage.

Perhaps we see an equation that looks like this:

$$x^2 = jabberwocky$$

At first, this might frighten us. How can we simplify that x^2? What manipulation can we perform on "jabberwocky?"

We know that the equation scale will not be tipped if we do the same thing to both sides. We can predict that the square root of x^2 will be x or -x. That might be handier to work with. If we take the square root of both sides, the equation will still be true:

$$\sqrt{x^2} = \sqrt{jabberwocky}$$

If we refer back to our handy chart on page 138, we see that the square root of jabberwocky is "slithy toves." The answer becomes apparent:

$$x = slithy\ toves$$

On the other hand, we are just as likely to see a problem that looks like this:

$$\sqrt{y} = slithy\ toves$$

We have no chart that indicates what the square root of "y" might be. Are we stuck? Not quite.

We know, without even thinking about it, that if we take the square root of something then square the result, we're back to our original number. The square root of 25, for example, is 5. Five squared gets us back to 25. This means we can square both sides of an equation without tipping the scales. Reversibility. Squaring our problem looks like this:

$$\left(\sqrt{y}\right)^2 = slithy \ toves^2$$

When we square the square root of y we get y. When we square slithy toves (refer to our chart, pg. 138) we get "jabberwocky."

So,

$$y = jabberwocky$$

Both squared numbers and square roots are game pieces with all the membership rights 23 or x have. This is easy to forget because they look so weird. But once you realize that it's true, you can do some interesting looking problems that are no more difficult than fourth grade arithmetic. For example:

$$x^2 + x^2 = ???$$

you would know the answer instinctively if x^2 means "1 apple." 1 apple, plus 1 apple equals 2 apples. Similarly:

$$x^2 + x^2 = 2x^2.$$

Another example:

$$\frac{\sqrt{xyz}}{\sqrt{xyz}} = ???$$

Einstein would cower at such ugliness. Until he realized that he had a fraction with the same numerator and denominator. It's just like 4/4, which equals 1. And this formidable fraction also equals 1.

More Tenor Flogging*

Flogging tenors is an activity. In algebra, there are four activities: adding, subtracting, multiplying, and dividing. An activity is called an "operation."

Let us represent flogging with a little diamond, like this: ◊

Kenn will be the flogger, and the tenors will be the floggees. We can abbreviate Kenn with a K. We'll abbreviate the tenors as T.

Now our little game looks like this:

$$K◊(T)$$

We can replace "tenors" with its equivalent. There are three tenors in the group. As a mathematician might say,

$$K◊(T)=K◊(B+D+S)$$

assuming that Bill, Dwight, and Steve are the tenors and have agreed to be abbreviated.

The distributive law applies to flogging That is:

$$K◊(B+D+S) = (K◊B)+(K◊D)+(K◊S)$$

Our intent was to flog all the tenors. We've certainly done that, and enjoyed it.

If we tried to group things differently, we might not get the job done.

$$K◊(B+D+S) \text{ does not equal } K◊B +D+S$$

Written that way, Bill gets flogged, but Dwight

and Steve escape. Not our intent at all.

But, perhaps those tenors have really been singing loud, and a simple flogging by Kenn does not convince them to mend their ways. Perhaps they have become calloused to the whole thing. Clearly, they will have to be flogged by both Kenn and Dr. Jim.*

$$(K+J)◊(T)$$

Kenn will flog each tenor, then Dr. Jim will flog each tenor:

$$(K+J)◊(B+D+S) =$$
$$(K◊B)+(K◊D)+(K◊S)+(J◊B)+(J◊D)+J◊S)$$

If our activity is multiplication, rather than flogging, a similar problem might look like this:

$$(a+b)(x+y+z)$$

The answer will be the same:

$$(a + b)(x + y + z) = ax + ay + az + bx + by + bz$$

Every floggee gets flogged by every flogger. It's been a good day.

Editor's Note: Kenn sings baritone, the common-sense part. Flogging is for illustration only. We do not advocate violence. Usually, merely withholding a tenor's sparkling water is enough.

Multiplying Polynomials

Multiplying polynomials is like flogging tenors, although it's rarely as much fun. Given the challenge:

(Jim + Kenn) flog (Bill + Dwight + Steve)

Jim flogs each tenor in turn, then Kenn does:

$(K+J)\Diamond(B+D+S) = K\Diamond B + K\Diamond D+ K\Diamond S +J\Diamond B+J\Diamond D+J\Diamond S$

If we substitute real numbers, and our activity is multiplication, rather than flogging, the same process looks like this:

$$(2+3)(4+5+6) = 8+10+12+12+15+18 = 75$$

To check our method, we might try doing the arithmetic within the parentheses first:

$$(2+3)(4+5+6) = (5)(15) = 75$$

This gets us the same number. When each polynomial can be expressed as a simple number, like this, your instincts tell you to do it the simplest way. And your instincts are correct. Why transform a simple problem into a more difficult one? You don't. You only use Little Weenie Numbers to understand how the process works.

In the tenor flogging example, every single term was a letter. In that case, you can't really apply any arithmetic to it. Most problems involve both numbers and

Editor's note: Dr. Jim also sings baritone.

letters, like this:

$$(2x+3)(x-2)= \ ??$$

The first term, the "2x" (now playing the part of Kenn) must flog each term of the second polynomial:

$$2x \cdot x = 2x^2$$
and
$$2x \cdot (-2) = (-4x)$$

Then the second term (in the role of Dr. Jim) flogs them all again:
$$3 \cdot x = 3x$$
and
$$3 \cdot (-2) = -6$$

We corral all our answers and sit them down together:

$$2x^2 \quad (-4x) \quad 3x \quad (-6)$$

Then we combine the results. That means adding them up. As a practical matter, it means putting a plus sign between our answers. We can only combine apples with apples and sopranos with sopranos, however. We can only combine "like terms." In this case, we have three varieties of results. We've got a plain old number, namely (-6). We've also got a certain number of "x's." We've got three positive ones and four negative ones. Those we can combine and we'll get (-1x). We've also got two "x squared's." If we had any other "x squared's" we could combine them as well. But we don't so our answer is

$$2x^2 + (-1x) + (-6)$$

153

We can clean this up a little. Adding a -1x is the same as subtracting 1x. And, while we're at it, 1x is the same as x, so we don't really need to bother including the "1." Our answer will look like this;

$$2x^2 - x - 6$$

Patterns

Life is full of repeating patterns. We learn where the store keeps our favorite potato chips, and what time the football game starts. If we didn't recognize hundreds of patterns, we'd spend all our time searching grocery stores shelf by shelf, item by item. We'd never see a football game, or find a bus stop.

Many people have trouble learning things without identifying the patterns. Until they realize that ostriches always hatch from ostrich eggs, and robins from robin eggs, matching egg to chick is a nightmare of trial and error. Math is often especially confusing to these folks. While more easily-led students cheerfully memorize rules and theorems without protest (and wind up with the good jobs) more critical souls may resist. They must "understand" why things work. Their math book won't tell them, so they'll feel confused and frustrated. Ultimately they will spend their lives as brilliant, if destitute, poets, philosophers, and social workers.

We have good news for these people.

Some kinds of algebra problems show up over and over. You may choose to beat your head against these problems for hours trying various strategies until they break down and surrender to your logic. This is challenging and aerobic, but not especially efficient. Once you've convinced yourself that the book and teacher aren't lying to you, that the examples make sense, the smart tactic is to simply remember them. Some students do this instinctively. They don't try to establish a historical and political context for the signing of the Magna Carta. They memorize the date, 1215 AD, and get on with their lives.

But many of us resist this common-sense approach. We insist on understanding the philosophical under-trimmings of squaring a binomial before we feel comfortable moving on to the next concept. Because neither your book nor your instructor wants to wait for your understanding to blossom, you will soon find yourself floundering in cold, deep water, feeling confused and stupid.

It's not your fault, however. They should make sure you understand this from the beginning: You can recognize several types of problems just by their appearance. And once you recognize them, you'll know the way the answer will look. In fact, some are so common and easy to remember, that you'll be able to solve intimidating equations in your head. For example, you will soon see:

$$(x+3)(x-2)$$

and know the answer faster than you can write down the problem.

Your algebra book will probably devote a section to each of these common problems. You'll see sections called "binomial squares," or "factoring the difference of squares." The idea of each of these sections is to illustrate a very common type of problem, so that you'll recognize it later, and automatically know how to solve it. Your teacher will believe this is obvious. The author of your Real Algebra Book will think this is obvious. For many of us, it's not at all obvious.

Beginning fishermen throw their lines into any body of water, at any time of day or night, with bait or lure chosen nearly at random. Over time, experience teaches them that certain catfish cruise the bottom and feed at night, while trout feed near the surface at dawn and dusk. The bait a mountain trout craves may insult any self-respecting catfish. Experienced fishermen put the right enticement on their line and drop it into promising waters at the right time of day.

156

Similarly, when fishing for the answer to some equation, beginners try things at random. But there are only a few types of problems repeated in variations over and over again. Instructors assume that sooner or later you'll recognize a binomial squared, or the product of two binomials involving addition. Once you do, solving them won't require any thinking. It won't require much time, or complicated arithmetic. You'll know what the answer will look like, and you'll simply plug the numbers into the picture. The sooner you grasp this idea, the likelier you are to enjoy algebra.

For example: Polynomials, when multiplied, emerge as creatures with distinctive markings. Every algebra book will tell you this, but they don't emphasize it. You'll read it and nod your head, but you won't take it to heart. You'll solve one equation, then treat the next one as a brand new adventure, completely unrelated to that last one. You will waste much time trying to corral squawking chicks before the concept sinks in.

If we multiply a binomial by itself, the answer will always look very similar to every other binomial squared. If we multiply it by another binomial involving addition, it will look almost exactly like every other problem like that. If the second binomial involves subtraction, with practice, we'll recognize it at a glance. If we multiply it by a trinomial, again, the result will have a distinctive appearance.

If you're one of those folks who can memorize such things easily, you probably aren't reading this book. You're deep into calculus by now. For the rest of us, this concept becomes valuable once we begin practicing on live equations. Sooner or later you'll start to recognize the promising fishing holes, and ply those waters. You won't waste time casting earthworms into the trees, dangling bait in your lunch pail, and trolling through the sand dunes.

For Example, Binomials Squared

There's no special trick to squaring binomials. It's just another polynomial multiplication problem, and we've already talked about that. What we're looking for now is the specific appearance of the result. When we square the binomial (*apples plus bananas*), the exercise looks like this:

$$(apple + banana)(apple + banana)$$

By now you can probably do this in your head. We multiply the first term in the first polynomial times each term in the second polynomial. Then we multiply the second term in the first polynomial by each term in the second. Just like flogging tenors. The results of each of these multiplications looks like this:

apple times apple	=	$apple^2$
apple times banana	=	*apple(banana)*
banana times apple	=	*banana(apple)*
banana times banana	=	$banana^2$

Can we simplify this at all? Are there any like terms?

Yes. *Apple* times *banana* is the same as *banana* times *apple*. We have two of these. So the answer will look like this:

$$apple^2 + 2(apple)banana + banana^2$$

If we abbreviate it, the problem would look like this:

$$(a+b)^2 = a^2 + 2ab + b^2$$

If the binomial involves subtraction, rather than addition, the answer will always look like this

$$(a-b)^2 = a^2 - 2ab + b^2$$

If we want to test our results against reality or common sense, we employ Little Weenie Numbers. Let's square the binomial of 2 plus 3:

$$(2+3)^2$$

We'll arrange it like a polynomial multiplication problem, which is, of course, its true identity:

$$(2+3)(2+3)$$

Now we'll multiply the first term, which is 2, by each of the terms of the second polynomial:

$$2 \cdot 2 = 4$$
$$2 \cdot 3 = 6$$

Then we'll multiply the second term of that first polynomial, which is 3, times each member of the second polynomial:

$$3 \cdot 2 = 6$$
$$3 \cdot 3 = 9$$

The results of all these multiplications are 4, 6, 6, and 9. When we combine "like terms" (which all of these happen to be) we get 25. Then we look back at our original problem, which is to square the sum of 2 plus 3. Five squared is 25.

How does that compare with simply plugging in our binomial squared recipe?

$$(a+b)^2 = a^2 + 2ab + b^2$$

If the result is going to look the same as our banana problem, it should look like this:

$$(2+3)^2 = 2^2 + 2(2 \cdot 3) + 3^2$$

We complete the arithmetic in the answer:
    2 squared equals 4;
    2 times 3 equals 6, times 2 equals 12;
    3 squared equals 9;

Adding these up, 4 plus 12 plus 9 equals 25, which we've already decided was the answer.

Perhaps this doesn't look much like the apple-banana problem, but it is. If we square a more threatening binomial, like $(xyz + 345)$ we already know what the answer will be:

$$(xyz + 345)^2 = (xyz)^2 + 2(xyz \cdot 345) + 345^2$$

We can get even more exact by completing the arithmetic. But that's for your teacher to worry about.

The interesting and useful concept here is that you're going to be squaring a ton of binomials before they let loose of you. If you get it into your head that the answer will always look pretty much like this, but with different numbers and variables plugged in, you will spend much less time chasing ostriches.

There are half a dozen types of polynomial multiplication problems that you'll see repeated in endless variations, between now and when you finish graduate

160

school. If you learn to recognize them, you'll know how to solve them automatically. No thought required. Just the way we like it.

FOIL

Multiplying polynomials is like flogging tenors. First, one member of the flogging team flogs each tenor, then the next member of the team does so. Finally, you combine like terms and take a well-deserved rest.

When you multiply polynomials times each other, you multiply each term of the first polynomial times each term of the second polynomial, combine like terms, and take a well-deserved rest. This system always works, regardless of how many terms each polynomial contains.

A binomial is a creature of two terms, like $(x + 1)$. You're going to spend a lot of time multiplying binomials times each other. The principle described above works just fine. You'll have four little multiplication problems, then you'll combine like terms. In this example:

$$(x + 2)(A - 3) = ???$$

The multiplication problems are these:

$$x \bullet A = xA$$
$$x \bullet (-3) = -3x$$
$$2 \bullet A = 2A$$
$$2 \bullet (-3) = (-6)$$

There aren't any "like terms" so you're done.

$$(x + 2)(A - 3) = xA - 3x + 2A - 6$$

To remind you to complete all four multiplication problems, someone invented the "FOIL" method. The letters in FOIL stand for **F**irst, **O**utside, **I**nside, **L**ast, which abbreviates the four multiplications you have to remember when multiplying two binomials. That is, in the above example, you multiply the two first terms, x and A; then the outside terms, x and (-3); then the two inside terms, 2 and A; and finally the two terms that come last in each binomial, the 2 and the (-3). Then you combine like terms, and you're done.

Nothing's unique about multiplying binomials compared to any other polynomial. The FOIL method doesn't imply any special properties or use any fancy ideas. It's just a handy reminder of the four multiplication problems.

Dr. Jim Calmly Discusses FOIL

A skinny box in my kitchen drawer says:

Aluminum Foil - Strong - Flexible.

The trouble is that most math students who learn to use FOIL don't remain flexible. They memorize "FOIL" and forget why it works.

162

Gerald, the second chair saxophone player, learns FOIL. He gets through the problems quickly so he can go out with that tall blonde baton-twirler, Geraldine. Let's listen in as he works the problem:

Multiply: $(x + 2)(4x + 6)$

"OK, gotta hurry here. . . This looks like a FOIL problem. Good! OK "FIRST" ah . . . that's the x times the $4x$. OK that's $4x^2$. Now "OUTER" ah . . . that's the x times the $+6$: OK that's $6x$. Now, ah . . . oh yeah. "INNER" ah . . . that's the $+2$ times the $4x$: OK that's $8x$. Whew! Now, . . .ah . . . "LAST" OK. That's the $+2$ times the $+6$: OK that's 12. So, the answer is $4x^2 + 6x + 8x + 12$. Combining like terms, I get $4x^2 + 14x + 12$."

In contrast, let's listen to how Geraldine talks her way through the problem. (Note that her work takes no longer to do!)

Multiply: $(x + 2)(4x + 6)$

"OK . . . this is a multiplication problem. So, I need to multiply every term in the first factor times every term in the second factor. So I'll take the x times the $4x$. OK that's $4x^2$. Now I'll multiply x times the $+6$. OK that's $6x$. Now I've done x times every term in the second factor. So, now I'll multiply $+2$ times the $4x$. OK that's $8x$. Next, I'll multiply the $+2$ times the $+6$. OK that's 12. So, the answer is $4x^2 + 6x + 8x + 12$. Combining like terms, I get $4x^2 + 14x + 12$."

What happens to this couple when they encounter this problem?

Multiply: $(x + 2)(x^2 - 4x + 6)$.

Because it doesn't look like a FOIL problem Gerald will be stuck. He'll get upset because he can't "remember" what to do now. He memorized meaningless steps and has no back-up system. If he tries to use FOIL, it will give the wrong answer. On the other hand, Geraldine's system will lead her though the problem correctly.

She learned to do the problem via the underlying principle, namely, the distributive principle. Because many students don't want to really understand, they fall for the tricks that make it look like they can do algebra. In the long run, they get penalized. In the short run, they get to go out with Geraldine.

Deciding to study a subject by learning the basic principles and underlying concepts is very important decision. It's a "third-little-pig" type decision — build the brick house. It is like saving 10 per cent of every paycheck.

Some people use a hybrid of these techniques. For example, they might use FOIL to get started. But they remember that they are faking it. They really don't understand it. They would not go on until they find the basic principle lurking behind the scene. In band, these folks are typically the lower brass players; trombone, tuba, etc.*

Knowing that you are faking some parts of algebra is the first step. At least, you are aware that something is missing. The next step is to look for the big picture. Don't be satisfied until you have the principle in mind. Teachers can sometimes help here.

You won't find the big idea by working lots of exercises. That's the wrong thing to do if it keeps you from thinking about the ideas. Many students of algebra would rather do extra exercises than think about the ones they have done.

It is much easier to remember a few underlying ideas than it is to remember thousands of separate problems.

Editor's Note: Dr. Jim plays tuba.

Factoring

Sometimes, the simple way to say something isn't perfectly accurate in all situations. Rather than be perfectly clear, books tend to be perfectly accurate, which is great, if you can just figure out what in the world their author is trying to say.

When we multiply two numbers, we call each of the numbers "factors." In the problem:

$$3 \cdot 2 = 6$$

both "3" and "2" are factors. The answer, "6," is the "product."

No problem.

In the multiplication problem:

$$123 \cdot 16x = 456$$

we can locate the factors easily. Both 123 and 16x are factors. They're the two "numbers" we multiplied times each other to get the "product" of 456.

Your math teacher may squawk at that last paragraph. 16x isn't really a "number." It's a monomial. To allow "16x" to be a factor, we have to expand our original definition to include "monomials." Once we let monomials play, we've opened the door to other polynomials, exponents, and the whole circus of other weird numbers, letters, and nonsense syllables. Which is exactly what we do. Any items you multiply together are factors. That's not difficult. But when you try to express that simple idea to someone who has no clue what you're talking about, and you're concerned with being scrupulously accurate, you wind up sounding like a lawyer arguing for custody of his polynomials. Things you multiply together are factors. Sometimes they're numbers, sometimes they're unknowns, sometimes they're fractions, or polynomials.

If you multiply a binomial times another binomial, like this:

$$(x + 1)(x - 3) = ???$$

each binomial is a "factor." In this case, "$x + 1$" is one factor, and "$x - 3$" is the other factor.

In arithmetic, usually we started out with ingredients and tried to make cookies. In algebra, sometimes we start with the cookie and try to deduce the recipe. When we begin with the "product" and try to figure out the multiplication problem that led to it, we are engaging in "factoring." The fun (and challenging) aspect of factoring is that there may be several correct answers. Or there may be only one. Or none.

Factoring means "expressing a number (or polynomial, etc., etc.) as a multiplication problem." Just as we've seen advantages to expressing "1" in various ways, there are sometimes tactical advantages to expressing things as multiplication problems.

We could express "12" as "3 times 4" for example. We could express it as "2 times 6" or "12 times 1" and we won't have affected how much pizza we've got, or how many tenors.

If you are told to "factor 12," then any of these answers would be correct. But one might be more useful to you, strategically, once you're using factoring as a weapon. Until then, you'll be given many expressions and commanded to "factor" them. Your classmates will dutifully transform these into multiplication problems, with no clue why they'd ever want to. You, however will understand. You have read this book.

Consider the number "16." If you were asked to factor it, what might you come up with?

Your answer will be a multiplication problem whose answer is 16. You might say, "8 times 2." That would be a correct answer. You might say "16 times 1." That would also be correct.

You don't need to restrict your answer to multiplication problems with only two factors. You might say "4 times 2 times 2." Or, your answer might look like this:

$$16 = (2)(2)(2)(2)$$

When you add an unknown to the problem, things become more interesting in this way: you'll get wrong answers for a while. When you factor "16x" for example, perhaps you'll try $(2x)(8x)$. This is the same as $(2)(x)(8)(x)$ It doesn't matter what order we multiply things, if that's all we're doing. So we rearrange these and get $(x)(x)(2)(8)$. First thing we come up with is x^2. And our original answer didn't have an x^2. So we miss that problem.

At that point, it will dawn on you like sunrise on Hawaii: only one of the factors must include an "x." Because if they both do, the product will have an "x^2." Possible answers will be things like:

$$16x = (2)(8x)$$
$$16x = (2)(2)(2)(2x)$$
$$16x = (4x)(4)$$
$$16x = (2x)(8)$$

The game becomes intriguing as the products become more complex. Perhaps we are told to factor "16x + 20." We are really being asked: What multiplication problem would result in that? First, we write it out:

16x + 20

Well, it's going to have an *x* in it. That much we know. But not two of them. Beyond that, we will stare at the page for a long time. It will seem impossible. In desperation, we'll try to think of any possible relationship in any multiplication problem that might result in a 16 and a 20. It occurs to us that "4" is a factor of both numbers. In fact, "4" times "4 *x*" would actually result in "16 *x*." And 4 times 5 makes 20. This insight ought to help us. We just need to think of a way to organize it.

After staring at the page for several more minutes, perhaps mowing the lawn and watching a basketball game in between, we might come up with this:

$$4(4x + 5)$$

If we perform the arithmetic, we're back where we started, which makes us happy. We have factored a binomial.

At this point, there is good news. Many, many binomials, when factored, will have a shape like this. One factor lurks outside a parentheses, a binomial cowers within it. The one outside is called a "common factor." It's common to both members of your original binomial. It divides evenly into each of them.

We might try factoring a binomial constructed of Little Weenie Numbers:

$$4 + 6$$

After our last experience, we look for clues. We want a common factor, something that "goes into" both 4 and 6. After some meditation, we realize that "2" goes into both of these. That's likely to show up outside our parentheses, like this:

$$2(?? + ??)$$

Little Weenie Numbers make us feel powerful one more time, as the answer springs from our brain. Two times 2 is 4, 2 times 3 is 6. So, one factored version of "4 + 6" is

$$2(2 + 3)$$

We have transformed "4+6" into a multiplication problem. Either way you complete the arithmetic, you haven't changed the size of the orchestra, or the number of push-ups the sergeant will require of you. In the real world, both descriptions represent "10."

The danger of using Little Weenie Numbers is that your intuition may lead you to a correct answer before you've got any idea what you did. To feel properly empowered, you need to retrace your steps when this happens, and decide what procedure your mind did without informing you. In this case, once you had the common factor, your brain probably started doing division problems.

Although factoring problems often look like escapees from a Stephen King novel, a few techniques will subdue most of them. If you can identify a common factor that's a monomial, (like "2" was in our last example) you're halfway home. You can factor by grouping. There are some specific tricks you can perform with exponents. And there are a few patterns that repeat so often you will learn to recognize them nearly automatically.

But some cannot be factored. They simply can't be transformed into a legitimate multiplication problem that, when completed, gets you back to the original. Perhaps there are no common factors.

Therefore, heed this warning: Factoring is a powerful weapon, once you know how to load it up and fire it. But it's not always the best weapon for the situation. So, while you're trying to learn to operate it, you will be given many opportunities to factor things that simply can't be factored. In those cases, the correct answer is, "Can't be factored." This isn't because math writers are cruel. At least, that's not the only reason. It's because they want you to have experience in deciding when factoring is an inappropriate strategy, later on. If you successfully factor everything they throw at you in basic training, you'll trust it too much when they start using real bullets.

Why They Make You Factor

Factoring is, on the surface, a foolish waste of time. You will read many books and factor many polynomials before you have a clue why you are doing it. The next sentence in this book will save you two months of confusion.

We factor to take advantage of some neat properties of zero.

In addition and subtraction problems, zero is powerless. You can add or subtract zero to a number for the rest of your life and you won't change the number at all. But zero becomes an all-powerful super-hero once you move into the domain of multiplication.

When you multiply any number by zero, you get zero. A million times zero equals zero.

When you divide zero by any number, you get zero. Zero divided by a million equals zero.

But you can't multiply any other two numbers

together and get zero. And you can't divide any other number by anything and get zero. If the answer to a multiplication problem is zero, one of the numbers you multiplied must also be zero. And, if the answer to a division problem is zero, your original numerator was zero.

That's why we factor.

If you can transform an equation into a multiplication problem whose product is zero, you know that one of the numbers you multiplied was zero.

$$(Catbox + Deadbeat)(Howlers) = 0$$

Either *(Catbox + Deadbeat)* equals zero, or else *(Howlers)* equals zero. Without a clue as to what scene is being played out, we know that's the only way to get a product of zero out of any multiplication problem.

To solve the problem, we first assume that *(Catbox + Deadbeat)* equals zero, and see what that tells us. While we're doing that, we simply ignore *(Howlers)*.

Once we're done with that, we set it aside and assume *(Howlers)* equals zero, and see what result we get. While we're engrossed in that, we ignore the *Catbox* part of the equation.

In other words, once we transform an equation into a multiplication problem whose answer is zero, we stop working with the multiplication problem. Instead, knowing that one of those factors must be zero, we investigate each factor in turn. We assume it equals zero and see what that tells us.

Oddly enough, we may get more than one correct answer. Assuming that *(Catbox + Deadbeat)* equals zero gives us one viewpoint, assuming *(Howlers)* equals zero may give us a completely different one. If this were Real Life, that annoyance would drive us crazy. But it's not. One of them must equal zero, and we don't really care which one. We get two different answers, and we love them equally.

Of course, should you ever be forced to solve an equation by factoring in the real world, you'll have to summon up some common sense to determine which correct answer is actually most useful to your situation.

Once an equation has a zero alone on one side, we can perform some operations that seem almost like cheating. Dividing both sides by any number won't change the zero at all. Even if the problem looks intimidating:

$$10(x^{32})(x-3) = 0$$

We can divide both sides by 10, which will cancel out the 10 on the left. Because zero divided by 10 is still zero, it won't have any effect on that side of the equation at all. This makes some sense. Some member of that left side is a zero. But not the 10. 10 does not equal zero. If it were seven million instead of 10, we could still get rid of it. The zero hidden inside one of those parentheses is going to destroy it anyway. So:

$$(x^{32})(x-3) = 0$$

The contents of one of those parentheses equals zero. If it's the first one, we know what x stands for. It equals zero, because that's the only number we can square, or cube, or raise to the 32nd power that will equal zero. So zero is one of our possible answers.

If the contents of the second one equals zero, then

$$x-3 = 0$$

By adding three to both sides of the equation we

get this:

$$x = 3$$

Three is another good answer. You can substitute either a 3 for the *x*, or a zero, and both will work. Both 3 and zero are "roots" of this equation.

Common Factoring Patterns

In every sport, we learn to recognize repeating events. Football players soon learn that the opposing quarterback is likely to throw the ball to one of his buddies. The first time this happens, we stare in amazement as they catch it and run past us. Can they do that, we ask? Is that legal?

Good defensive players learn to recognize clues that this is about to happen: the quarterback drops back, holding the ball a certain way, scanning the field, while receivers streak toward the end zone, wave wildly at the quarterback, and scream, "I'm open! I'm open!"

The clues prepare the defender. They help him plot an opposing strategy. It is best if the defender prevents this fellow from catching the ball and carrying it past him to the goal line. Intercepting the pass is one acceptable strategy. Knocking the ball out of the air is another. Yelling something profound and compelling at the receiver may distract him. If he does catch it, he should be tackled. Failing that, the defender should point to a team-mate, generously sharing the blame.

One would need to be a genius to play football if every play was completely unique. Because each play is a variation on a very few possible patterns, brain power is rarely the first attribute football coaches look for in a player.

Factoring is a lot like that. Each new variation will bewilder you. But the quarterback has a limited number of options, and you have only a few strategies to respond with. There are only about six common factoring situations. Once you identify which one applies, your brain can go back to sleep while you complete the routine arithmetic.

That's the important reassurance your Real Algebra Book will fail to give you. Factoring problems, like cookies, come in many shapes and sizes. But they're created from a very few recipes.

We usually deal with monomials, binomials, and trinomials. Any of these can participate in multiplication problems; any of them can be factors.

First, consider the lowly monomial. A monomial is already factored. It's a multiplication problem, all by itself. 4xyz means "4 times x times y times z." You might write it like this:

$$(4)(x)(y)(z)$$

Or, you might group it differently:

$$(4x)(yz)$$

As long as multiplication is our only operation, it doesn't matter what order you perform it or how you group the factors. So, factoring a monomial is trivial, and no one will ever ask you to do it. It may be useful sometimes to rearrange it, however.

Second, consider exponents. An exponent indicates something has been multiplied by itself, perhaps several times. So, to factor a number with an exponent, you only have to restate the multiplication problem that the exponent abbreviates:

$$x^2 = (x)(x).$$

$$x^2 y^2 = (xx)(yy)$$

$$x^4 = (x^2)(x^2)$$

How many ways can we multiply monomials, binomials and trinomials times each other? There are six possible combinations. Each one will have variations, based on whether the polynomial involves addition or subtraction.

Each answer will assume a characteristic shape. Six is not a huge number. Basketball has more varieties of penalty than that, and many people seem to have no difficulty committing them all.

And you're not going to run into these equally often. You will deal with the simpler ones more often. The common multiplication problems are:

- a monomial times another monomial,
- a monomial times a binomial,
- a monomial times a trinomial,
- a binomial squared,
- a binomial times another binomial,
- a binomial times a trinomial

When you factor, your answer will probably be one of these multiplication problems, or several multiplication problems, like (3)(4)(7)(2).

The nastiest factoring problems you're likely to

encounter will be variations on one of these very predictable patterns. You can save lots of time if you simply
notice this while you're learning. A couple of examples
may show how this works. The problem may look like
this:

$$\text{"Factor } x^2 \text{"}$$

The shape of the cookie suggests that the recipe looked
like this: $(x)(x)$. That's an answer. So is $(-x)(-x)$. Perhaps you can guess how z^2 would look when factored, or
how XYZ^2 would look when factored. They would look
very similar to the one you just did. Another problem
might look like this:

$$\text{"Factor } x^6 \text{"}$$

You could state the multiplication problem that led to this
several ways, including these:

$$(x)(x)(x)(x)(x)(x)$$
$$\text{or}$$
$$(x^2)(x^4)$$

Each are fine answers. Another example: A binomial, when squared, acquires a certain look. Once you
recognize this "look" you will be able to easily factor the
product of a binomial that has been squared.

$$(a + b)(a + b) = a^2 + 2ab + b^2$$

The trick will be recognizing this relationship, this "look,"
when numbers and unknowns replace "a" and "b." Sooner
or later, your mind will click, you'll recognize the pattern
that this tries to depict. When it does, if you see something that looks like this:

$$x^2 + 2xy + y^2$$

you will guess that the multiplication problem that led to it was a binomial squared, and the binomial was probably "$x + y$."

Then you'll see something like this:

$$9x^2 + 24x + 16$$

and something magical will happen within you. You'll say, hey, that sure looks like a binomial squared! Let me check a little further here...why, yes! If the binomial was $(3x + 4)$ then squaring that gets me this answer!

I know you don't believe me. But it is the discovery of these patterns that separates the physicists and mathematicians from the accordion players.

Quadratic Formula

Almost 4,000 years ago, mathematicians in India and Iraq realized they could not solve all quadratic equations by factoring. But, by creative manipulation, they could solve some of them anyway.

They knew they could add the same thing to both sides of an equation. Perhaps by sheer luck, someone tried adding a fraction to both sides, a fraction that was not an obvious choice. With a couple of other neat tricks, they managed to solve their tough problems. It probably cost them a year of trial and error calculations, and a roomful of clay tablets. You can bet the next time they were given a tough quadratic problem, they immediately tried to find a similar strategy. Pretty soon, they discovered a definite pattern to the whole process. Once they identified the pattern, they realized that it worked every single time. It didn't matter if they could factor something or not. They could solve any quadratic equation.

By the time they had isolated the unknown on one side of the equation, the other side always looked similar to earlier problems. The numbers were different, of course, but the pattern was the same. So, why not skip the preliminary, disposable steps and save some time? Once they stated their quadratic equation in the "standard" way, they could simply translate it to their new "quadratic formula" and complete the arithmetic.

A "complete" quadratic equation in "standard form" will look something like this:

$$3x^2 + 5x + 12 = 0$$

The only unknown is "x." That's what we're trying to identify. The squared number comes first, the un-squared unknown is the second term, and a known number is the third term. This is the "standard form" of a quadratic equation. It has been "set to zero" and the numbers arranged a special way. That's an important concept. A complete quadratic equation in standard form will always resemble this, and we need to manipulate other quadratic equations until they do look like this, if we want to use the Quadratic Formula to solve them. So, stare blankly at that last one for a moment, in the vain hope that you'll recognize it when it's disguised.

Beside the unknowns, each of the *numbers* plays a specific role in this equation. Three is the coefficient of the squared unknown. 5 is the coefficient of the regular unknown. 12 is the plain old number. Even if the numbers were much larger, they would play the same role. Even if they were fractions, binomials, or other weird birds, their relationship to the equation is the same. The ancient mathematicians, joyful at discovering a huge short-cut, wanted to describe the roles of each of these numbers, so they could translate any other number into the appropriate role. What has been passed down are three logical, if not very colorful names: a, b, and c.

In the equation we described:

$$3x^2 + 5x + 12 = 0$$

Three plays the role of "a," 5 plays the role of "b," and 12 plays the role of "c." The Quadratic Formula is going to use the letters a, b, and c to tell you what to do. You will have to identify the numbers in your own problem that correspond to these roles. This is the "standard form." You need to manipulate quadratic equations until they are in this form to use the formula. When the problems look very similar to this one, you won't have any trouble. You'll see:

$$16x^2 + 32x + 127 = 0$$

and you'll immediately translate:

$$a = 16$$
$$b = 32$$
$$c = 127$$

Then you'll replace the a's, b's, and c's in the formula with those numbers, complete the arithmetic, and have the answer.

It becomes more challenging when the equations don't look identical to our sample. For example

$$x^2 + 5x + 12 = 0$$

You start to translate, and you stop cold. What happened to "a"?

Then you'll remember that the coefficient of x^2 is really 1. You always assume that, but rarely think about it. In this case, we have $1x^2$, and "a" is played by the number 1.

Once you remember that, you won't have trouble with something like this, either:

$$3x^2 + x + 12 = 0$$

The unstated coefficient of x is also 1, and that's what you'll use for "b."

If there is no "b" or "c," the problem will probably yield to simpler tactics. If nothing's squared, it's a linear equation. If anything in the equation is raised to a larger power than 2 (like x^4) it's not a quadratic equation, and the formula won't work. Unless you can ingeniously transform it into a quadratic equation, but that comes later.

The Quadratic Formula is a bit intimidating to gaze upon:

$$x = \frac{-b \pm \sqrt{b^2 - 4ac}}{2a}$$

But it's just arithmetic. It will always look just like this. It's the only really wild formula you'll have to memorize your first year, and about the wildest you're likely to ever encounter. Here's how it works.

You've been given a problem that looks like this:

$$rats^2 + (pigs)rats = doggy$$

Your first problem is to recognize that this has the makings of a quadratic equation in standard form. You get it there by subtracting doggy from each side:

$$rats^2 + (pigs)\ rats - doggy = \ doggy - doggy$$

which gets you to this:

$$rats^2 + (pigs)rats - doggy = 0$$

This has a certain familiar look to it, doesn't it? Yes. It's a lot like a quadratic equation in standard form. After careful examination, we realize that "rats" is playing the role of "*x*," and "pigs" is playing the role of "b." But the formula calls for us to ADD "c" and we subtracted doggy. Craftily we decide that we did not subtract a positive doggy. We added a "negative doggy" to play the role of "c." Because there's no coefficient in front of "rats squared," we know that "a" is "1." No problem. We simply plug these into the Quadratic Formula.

Here's the formula:

$$x = \frac{-b \pm \sqrt{b^2 - 4ac}}{2a}$$

And here's how it looks when we plug our information into it:

$$rats = \frac{-pigs \pm \sqrt{pigs^2 - 4(1)(-doggy)}}{2(1)}$$

If your calculator has a barnyard configuration, you can go ahead and complete the arithmetic.

Just kidding. This is only an example of how you identify which members of an equation play the various roles in a quadratic equation and how you put them into the formula.

When we factored, we discovered that there were

often two answers to a quadratic equation. Same thing happens here. The plus or minus sign means you try both adding and subtracting, and you'll get different answers.

Another little oddity is that negative numbers don't have square roots, in the traditional sense. You can't multiply any number by itself and get a negative number. But mathematicians wish they could. And, sometimes you're going to get negative numbers in the old Quadratic Formula right beneath that radical sign, causing yourself some confusion.

To ease your anguish, they invented a category of numbers called "imaginary numbers." The imaginary numbers are the square roots of negative numbers. The square roots of negative 4, are imaginary 2 and negative imaginary 2. Imaginary 2 squared is negative 4.

To see how neatly the Quadratic Formula works, let's make up a problem and take it for a test-drive. To avoid the stress of working on problems with unknown answers, let's make up an answer first. That way, when we plug it into the formula, we'll know for sure if we're right. We take a vote and decide our unknown, (x), will equal 4.

We'll start with

$$2x^2 + 3x$$

Just seeing that makes me uneasy, and I haven't even built a whole equation. It looks dangerous, I feel stress, so I cheat. Before adding the "c" portion, I casually calculate what I've got already. 4 squared is 16, times 2 is 32. So that first term is really 32. $3x$, (that is, 3 times 4) is 12. My second term is 12. So far, what I've got is 32 plus 12, which is 44. I'm going to call negative 44 my "c", so that when I add it, the whole mess will, indeed, equal zero.

$$2x^2 + 3x + (-44) = 0$$

Now we have a neat quadratic equation, whose solution we happen to know. Our unknown is 4.

Then we identify what numbers are playing the roles of a, b, and c. Because we have them all lined up properly in the "standard form," this is a cinch:

$$a = 2, \ b = 3, \text{ and } c = (-44).$$

Because we have not yet memorized the formula, we set it beside us and copy. We put a 2 everywhere the formula shows an "a," we put a 3 wherever the formula shows a "b" and we put (-44) every place the formula shows a "c." Then we complete the arithmetic, and, if we're lucky, we wind up proving that x equals 4. The form looks like this:

$$x = \frac{-b \pm \sqrt{b^2 - 4ac}}{2a}$$

When we substitute our own information, it will look like this:

$$x = \frac{-3 \pm \sqrt{(3)^2 - 4(2)(-44)}}{2(2)}$$

The next concern is, what order do we do the arithmetic? We can't do much with any of this until we pin down that square root. Within the radical sign (the square root sign) we first square the 3 to get 9. 4 times 2 is 8, times (- 44) is (-352).

$$x = \frac{-3 \pm \sqrt{9 - (-352)}}{2(2)}$$

Subtracting a negative 352 is the same as adding a positive 352. So the number under the radical sign is 361. You can look the square root of that up in our table, or you can use your calculator. Or you can experiment with numbers. Got to be bigger than 10 squared, because that's a hundred. Got to be smaller than 20, because 20 squared is 400. Turns out the square root of 361 is 19. If it had not worked out exactly, I'd be nervous, because I created the problem in the first place. After tidying up the other arithmetic, it looks like this:

$$x = \frac{-3 \pm 19}{4}$$

We have two more steps. We have to try adding negative 3 and 19, then try subtracting them. If we add a negative 3 to 19, we get 16:

$$x = \frac{16}{4} \text{ which equals four.}$$

If we subtract, we get:

$$x = \frac{-22}{4} \text{ which equals -5.5.}$$

Remarkably, a negative 5.5 works just as well in our original formula, as the "4" we designed it to work with. It's magic.

After you've got answers, it's smart to plug them into your original problem and see if they work. It's a built-in way to tell if you're right.

It's much more fun to create problems from scratch like this than it is to glue your head inside a boring textbook. You feel powerful, like an explorer, as you struggle with the same problems the ancient guys did when they invented this stuff. Your teacher may not mention that this is a reasonable, fun approach. The serious men who sell textbooks certainly won't. Despite the inevitable mistakes and frustrations, try it. If you are bold, and seri-

185

ously interested in owning algebra, that's how it will happen. You will also be able to work at your own pace, on the simplest problems you can invent, and cheat like a riverboat gambler whenever you want.

Little Weenie Numbers Demonstrate
The Invention of the Quadratic Formula

This chapter is completely unnecessary. It won't be on the test. It's just to make me feel better. If you get bored, skip it. I'll never know.

The first time I beheld the Quadratic Formula, I distrusted it. Where in heck did THAT come from, I asked? Why should I believe it's going to get me home from the prom without breaking down on some dismal back road?

If you also distrust evil-looking formulas,

described in words that you don't understand, by a professor with an undertaker's perky delivery, you may find comfort in Little Weenie Numbers. Return with us now, to those days of yesteryear, as we build "the quadratic formula" from the ground up.

As is my preference, I'll make up a simple problem. To reduce stress, I make up the answer first. x equals 2. We'll start with $x^2 + 3x$, and calculate that before we add "c." 2 squared is 4, plus 6 equals 10. So, let's make negative10 play the part of "c."

$$x^2 + 3x + (-10) = 0$$

We know this is true, because we already know the answer. This is just a test drive, to see what's under the hood. It's the formula being tested here, not us.

We quickly figure out who's playing each role. The unstated coefficient of x^2 is 1. So, "a" equals 1. The coefficient of x is 3. So, "b" equals 3. What will confuse us for a while is that the standard form calls for us to add "c." So, what we have to do is add a negative 10. Same thing as subtracting, of course, but if we don't keep track of that subtle difference now, while it's easy, when we're working as rocket scientists, we'll find ourselves landing on Neptune instead of in Cleveland. This is how it ought to look:

$$x^2 + 3x + (-10) = 0$$

To keep track of our Little Weenie Numbers, I'm going to keep each one in a neat little parentheses. Won't affect anything.

Here's the experiment: Let's replace x, a, b, and c, with the correct numbers, right from the beginning. Eliminate that unsightly unknown, just to see how the process works. We'll start out with a standard form equation, expressed both as familiar Little Weenie Numbers and as traditional letters. Then, we'll see if we can manipulate them, step by step, into the quadratic formula. It should be lots of fun.

So, our standard form, properly "coded." looks like this:

$$ax^2 + bx + c = 0$$

which we translate into Little Weenie Numbers:

$$(1)(2)^2 + (3)(2) + (-10) = 0$$

Treating this like a linear equation, we want to get all the x's (which are all "2" in our Little Weenie Number example) onto one side, and all the known quantities onto the other. First step is to subtract negative 10 from each side of the equation:

$$ax^2 + bx + c \ -c = 0 - c$$

or

$$(1)(2)^2 + (3)(2) + (-10) - (-10) = 0 - (-10)$$

This gets us to:

$$ax^2 + b(x) = (-c)$$

or

$$(1)(2)^2 + (3)(2) = 0 - (-10)$$

So far, we can do the arithmetic in our brains and be confident of our tactics.
Then we divide both sides by "a"

$$\frac{ax^2}{a} + \frac{b(x)}{a} = \frac{-c}{a}$$

or

$$\frac{(1)(2)^2}{1} + \frac{(3)(2)}{1} = \frac{-(-10)}{1}$$

Although we're not going to take advantage of each opportunity to simplify or perform arithmetic, in this case, we're going to cancel out those first "a's":

$$x^2 + \frac{bx}{a} = \frac{-c}{a}$$

or

$$2^2 + \frac{(3)(2)}{1} = \frac{-(-10)}{1}$$

Here's the part that required a little creativity and perhaps several quarts of Babylonian root-beer. We can add the same thing to both sides of the equation. What we're going to add, however, is a fraction, and not one you're likely to find lying around your camel shed. We're going to add

$$\frac{b^2}{4a^2}$$

to both sides of the equation:

$$x^2 + \frac{bx}{a} + \frac{b^2}{4a^2} = \frac{b^2}{4a^2} + \frac{-c}{a}$$

In our real number example, we add

$$\frac{3^2}{(4)(1)^2}$$

which is the same as adding nine-fourths to each side of the equation. Perfectly harmless, but very ugly. Now our Little Weenie Number example looks like this:

$$2^2 + \frac{(3)(2)}{1} + \frac{3^2}{(4)(1)^2} = \frac{-(-10)}{1} + \frac{3^2}{(4)(1)^2}$$

Again, we can quickly do the arithmetic and convince ourselves that everything's still equal.

We can combine the two fractions on the right side of our equation if we locate a common denominator. Because "a" goes evenly into "$4a^2$" we can use "$4a^2$" as the common denominator. We multiply the fraction by a variety of 1:

$$\frac{-c}{a} \bullet \frac{4a}{4a} = \frac{-4ac}{4a^2}$$

Quickly, before we panic, we compare this to our Little Weenie Numbers:

$$\frac{1(-10)}{1} \bullet \frac{(4)(1)}{(4)(1)} = \frac{-(4)(1)(-10)}{4(1)^2}$$

So, after combining these fractions on the right side of the equation, the equation looks like this:

$$x^2 + \frac{bx}{a} + \frac{b^2}{4a^2} = \frac{b^2 - 4ac}{4a^2}$$

which, in our trial example would be:

$$2^2 + \frac{(3)(2)}{1} + \frac{3^2}{(4)(1)^2} = \frac{(3)^2 - (4)(1)(-10)}{4(1)^2}$$

At this point, the Babylonian root beer inspired someone to notice something that I wouldn't see in a hundred years. The left side of the equation could be expressed as a binomial squared:

$$x^2 + \frac{bx}{a} + \frac{b^2}{4a^2} \text{ is the same as } \left(x + \frac{b}{2a}\right)^2$$

190

and, in our example,

$$2^2 + \frac{(3)(2)}{1} + \frac{3^2}{(4)(1)^2} \text{ is the same as } \left(2 + \frac{3}{(2)(1)}\right)^2$$

At this point, the complete formula looks like this:

$$\left(x + \frac{b}{2a}\right)^2 = \frac{b^2 - 4ac}{4a^2}$$

and our example looks like this:

$$\left(2 + \frac{3}{(2)(1)}\right)^2 = \frac{3^2 - (4)(1)(-10)}{4(1)^2}$$

By now you should be beginning to appreciate the joy these folks felt when they finally got something useful. Although there is some logic to these steps, none are inevitable. These are the steps that wound up working, after trying many other alternatives. You or I probably would have come up with them ourselves, given lots of time and desperately boring lives..

At any rate, to continue. The advantage of making the left member a square is that we can get rid of the square by taking the square root of both sides, leaving us with a little simpler mess on the left side, and a little uglier mess on the right side. Our goal, remember, is to have no mess at all on one side, regardless of the consequences for the other side. After taking the square roots of both sides, the equation looks like this:

$$x + \frac{b}{2a} = \pm \frac{\sqrt{b^2 - 4ac}}{2a}$$

In our example, we have this:

$$2 + \frac{3}{(2)(1)} = \pm \frac{\sqrt{3^2 - (4)(1)(-10)}}{2(1)}$$

The odd sign to the right of the equal sign is a "plus or minus" sign. Each problem will yield two different answers, the one we get when we add, and the one we get when we subtract. But then, we often get two different answers with quadratic equations.

Another quick note. The entire monster that is now within the radical sign is called the "discriminant." If it's a positive number, both solutions to the problem dwell in the land of real numbers. If it's a negative number, your solutions will be non-real. Because, of course, square roots of negative numbers are imaginary, but just as useful as other imaginary things. If the discriminant is zero, there's only one solution.

But I digress. We can clean up the equation even more by subtracting the fraction from both sides:

$$x + \frac{b}{2a} - \frac{b}{2a} = \pm \frac{\sqrt{b^2 - 4ac}}{2a} - \frac{b}{2a}$$

like this:

$$2 + \frac{3}{(2)(1)} - \frac{3}{(2)(1)} = \frac{\sqrt{3^2 - 4(1)(-10)}}{2(1)} - \frac{3}{(2)(1)}$$

After subtracting, it looks like this:

$$x = \pm \frac{\sqrt{b^2 - 4ac}}{2a} - \frac{b}{2a}$$

We also subtract in our example:

$$2 + \frac{3}{(2)(1)} - \frac{3}{(2)(1)} = \pm \frac{\sqrt{3^2 - (4)(1)(-10)}}{2(1)} - \frac{3}{(2)(1)}$$

To arrive at this:

$$2 = \pm \frac{\sqrt{(3)^2 - (4)(1)(-10)}}{2(1)} - \frac{(3)}{(2)(1)}$$

Now we combine that negative fraction way sitting off by itself with the rest of the right side. Because the denominators are identical, it's a simple subtraction problem, or adding of a negative number, or combining like terms. We could put "minus sign 3" after the radical above the single denominator, and that would be fine. In fact, I bet that's what they originally did. I bet they also wasted many days and much papyrus forgetting that the "3" wasn't part of the Nile Rat Stew inside the radical sign. So they put it to the left of the radical sign, and the finished version looks like this:

$$x = \frac{-b \pm \sqrt{b^2 - 4ac}}{2a}$$

And our Little Weenie Number version looks like this:

$$2 = \frac{(-3) \pm \sqrt{(3)^2 - (4)(1)(-10)}}{2(1)}$$

Are we convinced yet? If not, complete the arithmetic. The number under the radical sign is 49. The square root of 49 is 7. We have to try adding negative 3 to that, to get one root. Subtracting negative 3 from it will give us the other root.

Negative 3 plus 7 is 4. Divide that by 2, and you discover that 2 is one "root." (Remember: a "root" is a number. When you replace an unknown, like x, with a root, it makes an expression equal to zero. There may be more than one root for an expression.) Because we began by deciding that 2 would be our unknown, we feel proud and more confident of the entire process. But we still have to see what happens when we subtract, instead of add.

Negative 3 minus 7 is negative 10. Divided by 2 equals negative 5. Can this be right? Can "negative 5" be just as good an answer to our problem, even though we didn't expect it?

Absolutely. Try it in our original equation. But notice this: we started by knowing that x equals 2. And 2 does not equal negative 5. So, although either will work in the formula, one of these answers probably makes a lot more sense in the real world.

Lines and Points

A "line" is not that thing we draw with a stick in sand or the mark we make with a pencil on paper. No one has ever seen a line or drawn one. At least, not according to mathematicians.

To them, a "line" is an abstract beast with special, idealized characteristics. It resembles any line you might see or draw, with these exceptions:

1. It's perfect in every way.
2. It's endless.
3. It has no thickness whatsoever, and is therefore invisible.

Obviously, there are no perfect, invisible, endless lines outside the fertile imaginations of folks who spend too much time with their calculators. The lines we draw are crude approximations of this holy character. But all the laws of math presume that you're dealing with a perfect imaginary line. Real world calculations will always be limited by our ability to draw and measure. They will always be a tiny amount "off" because these perfect lines don't exist. Our finest pencil line will have some thickness and some irregularity.

A portion of a "line" is a "line segment." Line segments have a beginning and an end. They have a specific length. Except for that, they are just as perfect and imaginary as lines.

Once mathematicians start making up perfect abstract characters, it's hard for them to stop. A line has infinite length. A line segment has a very specific length. If you keep shortening a line segment until it has no length at all, you call it a "point." A point has no thickness, no height, no weight. It is simply a location. We often represent points with dots, like the period at the end of a sentence. But a dot of ink is not a point, any more than a pencil mark is a line. They are rough representations of imaginary, abstract ideas.

If you say that a line segment is a bunch of points strung together, no one will argue with you. They might not stand there and listen to you for very long, either, but they won't argue. A line is a series of points.

Graphing

Charts are part of our lives. We've all seen charts that illustrate how crime is growing or shrinking or how much money a company earned this year compared to previous years. It's easier to recognize patterns when they're displayed in a graphical way.

Most charts combine two different kinds of information. Going right, for example, may represent time passing, while moving vertically might indicate earnings.

Graphical illustrations describe patterns in an obvious way so we can quickly understand them and make predictions. They appeal to our primitive visual instincts. Because they can be understood without much thought, they are often used by politicians.

You don't have to study the charts on the next page very hard to decide which climate would be easier to get used to.

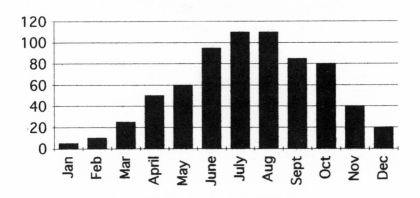

Average Temperature, Minnesota

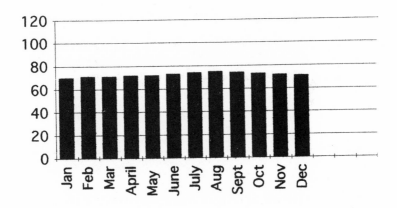

Average Temperature, San Diego

Editor's Note: These are not accurate charts. Minnesota has a wonderful climate. This is just another fun visual aid.

Algebra uses graphs in a similar way. They help you visualize problems and provide a different perspective when an equation stumps you. And they lead you directly into the calculus, analytical geometry, and trigonometry. If you gloss over the graphing portions of algebra, you'll be in a world of hurt if you ever try one of the more advanced math classes.

In 1637, about the time of the Pilgrims, a fellow named René Descartes made one of those wild leaps of the imagination that change the world. At this time, some people were fooling with algebra, using equations just the way you and I do. Other people enjoyed geometry, studying lines, triangles, circles, etc. But algebra and geometry were unrelated.

René noticed that the tops of the bars in our Minnesota climate chart could be connected in a line, or a series of lines. The line made by the top of our San Diego chart would look a lot different. Some charts might plot out triangles or other interesting shapes. He wondered if the shapes created by charts were subject to the laws of geometry. Was there any relationship between the two different activities?

As it turns out, if you set up the rules to your game carefully, the lines you draw to graph algebra problems also fit nicely into the geometry game. This revelation opened up a whole new world of possibilities and several new classes for math teachers to teach. The basic game plan for graphing algebra problems is still called the "Cartesian" method, after Descartes. Of course, he didn't really use our Minnesota climate chart. Minnesota had not yet been invented. We are told that he got his idea while watching a fly crawl from point to point on his ceiling. Of course, the only person we've heard that from is Joe Reid. Because Joe's last novel involves many flies doing unpleasant things, this may be more "subtle marketing ploy" than "historical fact." But it *could* have happened that way.

Math teachers may be unable to convey their enthusiasm for the Cartesian system enough for you to feel giddy and flushed along with them. But translating algebra into geometry, and vice versa, is powerful. It means that we can transform complex events into a chart and then use the relatively simple rules of geometry to explain and predict things. If you can plot the change in speed of a rocket as it moves between two planets, perhaps you can predict where it will go with your compass and ruler, rather than with eighteen pages of calculating equations. We can use graphs to create equations and equations to create graphs. We can look at things from two completely different perspectives, gaining new insights from each. This is a major strategic leap. It made possible space travel, nuclear reactors, video games and modern movie special effects.

But they won't tell you that—not right away, anyway.

First you must prove yourself Worthy by learning a whole truckload of jargon, then you must figure out the subtle differences between describing equations and graphs. For example, parentheses mean different things in equations and in graphs. An equal sign means something different. An "x" can now be either an unknown in an equation or a distance on a graph.

To heighten the fun, they won't spell out any of these little differences. They don't intend to be cruel. They don't even notice they're doing it. So you must ask questions when you feel confused.

It might be useful to hear a bunch of these new words, even briefly, so they don't seem quite so foreign during class. Let us swing quickly through the graphing jungle, swiping at the occasional juicy concept, to get a preliminary feeling for the terrain. Do not take notes. This is not a harvesting expedition, but an exploratory one.

━━━━━━━━━━━━━━━━━━━━━━

The Number Line

Graphing starts with the number line, which is a very long imaginary tape measure. At the center of the number line is zero, also called "the origin." Evenly spaced to the right of the origin we find consecutive numbers, starting with 1, 2, 3, etc., and continuing as long as you care to count. Evenly spaced to the left of the origin are negative numbers, beginning with negative 1, negative 2, negative 3, etc., also continuing until you have enough numbers for your purpose.

Any positive or negative number will be located somewhere along that line. Any fraction can be located. Conversely, any point on the line can be expressed as some number.

It only takes one number to describe any point on the number line. Two points describe each end of a line segment.

Sometimes a line represents a process that only goes in one direction. Time, for example, or gravity. When that occurs, you could put an arrow on the end of your line to remind you of this. Directional lines like that are called "vectors."

If you have two numbers, they'll both be on the number line, and the distance between them will be a straight line of some specific length. We can use a number line to express arithmetic concepts like addition and subtraction. 2 plus 2 might be a 2-inch section plus another 2-inch section, giving us a line 4 inches long. If you're having difficulty picturing some addition problem, you can always plot it out on the old number line.

Adding and subtracting affect the end of the line farthest away from zero, making it longer or shorter. If you're really talking about a single point on the line, then adding and subtracting moves the point to the right and left. This is one of the first subtle stumbling blocks to watch out for. Sometimes you'll be concerned with a single point, sometimes with a line segment.

Negative numbers move contrary to positive numbers. If adding a positive 2 stretches a line 2 units to the right, adding a negative 2 must pull the end of the line to the left. If subtracting 2 pulls a line closer to zero, then subtracting negative 2 must stretch it farther, making it a longer line.

Absolute Values

The distance from Albuquerque to Winnipeg is 1,492 miles. So is the distance from Winnipeg to Albuquerque. Distance in either direction is always described with a positive number. So is the length of a segment or vector.

It's easy to find the distance between two numbers on the number line: you just subtract them.

Well, almost. If you subtract them in the wrong order, you get a negative answer. But people don't like distances to be negative, unless they are physics majors. So mathematicians invented the absolute value signs to clean up the mess. When you see these vertical lines " | | ", you know the writer was thinking about distance.

Absolute values ignore the positive and negative aspects of a number. The absolute value of negative 13 is 13. We'd write that

$$|-13| = 13$$

The absolute value of positive 64 is 64:

$$|64| = 64$$

Both of these symbols ask how long a number is, ignoring its direction. Sixty-four has length 64. Negative 13 has length 13.

You subtract two numbers to find the distance between them. But if you subtract them in the wrong order, you get a negative answer. These little lines help prevent that mistake. That's their complete job description. To find the distance between 3 and 5 you can use either

$$|3 - 5| = 2$$

or

$$|5 - 3| = 2$$

The absolute value signs make the answer into a positive number if it isn't already. The distance between 5 and -8 is 13. You could write it as

$$|5 - (-8)| = |5 + 8| = |13| = 13$$

Or if you subtract them in the other order, you'd get

$$|(-8) - 5| = |-13| = 13$$

These little symbols "| |" cause trouble in algebra. While the algebraic steps are somewhat complicated, the intention of absolute value is easy. You don't even want to see how your Real Algebra Book will explain |x|. But when you have to read it, remember that distances are always positive. It'll help.

Graphing Points

A graph usually has both a horizontal reference, like the number line, and a vertical reference. The vertical line intersects the number line at the origin, which is zero. The vertical line is simply another number line, with positive numbers going up from zero toward the top of the page, and negative numbers going down from zero toward the bottom of the page.

Our original horizontal number line is called the "x axis." The vertical line is called the "y axis." Graphing is comparing points and lines to these two reference lines.

The two axes are like little rulers. Zero is always the starting spot. You can easily locate a spot three inches to the right of zero and 2 inches above it. First you measure three inches to the right, then 2 inches up from that and make a mark. That's the idea of graphing.

Or, we could make one mark three inches to the right of zero on the x axis and another mark 2 inches above zero on the y axis. Then, we'd draw a horizontal line through the mark on the y axis, and a vertical line through the mark on the x axis. The two lines would intersect at our destination. If we were using graph paper, with little blue horizontal and vertical lines already printed, we wouldn't have to draw these lines ourselves.

The first mark we made, the spot on the x axis, is called the "abscissa." The word comes from a Latin word, and translates loosely as a "cut-off line." The abscissa is the place where we "break off" the x axis. It tells us how far to the left or right of zero our point will be.

The second mark we made, the point on the y axis, is called the "ordinate." This word comes from a Latin word that means "to regulate," or "to organize." The "ordinate" on the y axis tells us how far above or below zero our point is going to be. The words "abscissa" and "ordinate" were concocted by Gottfried Leibniz, a German diplomat who was born about ten years after Descartes had his big idea. You'll hear more about him in calculus class.

Perhaps not surprisingly, you will begin by plotting points, then you'll plot straight lines, then curved lines, and in a future class, you'll be plotting triangles, circles, and strategies.

The numbers on a number line can represent any kind of unit. They can be miles, or inches, or pounds, or potatoes. Or, they might just be numbers, without any other significance in the real world. Sometimes they'll be referred to as "units."

To locate a point requires two distances. The first one will always be the distance from zero on the "x axis," which is the horizontal reference line. The second number will be the distance from zero on the "y axis," the vertical distance. They will surround these two distances with parentheses and call them an "ordered pair." They'll be separated by a comma. Because the x distance always comes first, they don't label it any other way. So, when you see (2,3) you'll know it describes a point 2 units to the right of zero on the horizontal line, and 3 units above that spot. The comma tips you off that it's an ordered pair, rather than a polynomial or group of tenors.

On a graph, negative numbers go to the left of zero (on the x axis) and below zero (on the y axis). The ordered pair (-3,4) locates a point 3 units to the left of zero on the x axis, and 4 units above that.

If a point isn't exactly on one of the number lines, it can be in one of four areas. It can be above and to the right of zero, above and to the left of zero, below and to the left of zero, or below and to the right of zero. Each of these areas is called a "quadrant." If your graph was a map, quadrant 1 is north and east of zero, quadrant 2 is north and west of zero, quadrant 3 is south and west of zero, and quadrant 4 is south and east of zero. They always label the four quadrants in this rather peculiar counter clockwise way. You'll be given ordered pairs, for example, and asked to tell in which quadrant the point they describe is located. The quadrants are identified with Roman numerals. You'll learn to love quadrants in trigonometry.

Graphing Lines

Straight lines are the simplest thing to graph after you have figured out how points are located. They are described by equations that don't seem to care what "x" is or what "y" is. They only care how a change in horizontal distance affects the vertical distance. Things that look like perfectly good equations include nothing to solve, no mystery to unravel. That's OK. We're playing a different game now.

For example, if you claim that

$$x = y$$

then every time you plot a point, it will be the same distance to the right or left of zero as it is above or below zero. If you go ten units to the right, you must go ten units up. If $x = (-6)$ then y also equals (-6). When you connect all the possible points, you'll get a straight line. You can plot points until your pig speaks French, and all the points will be somewhere on that one straight line described by $x = y$.

If y = x +3, your line will look a little different. Trying different possibilities at random, if x is 10, then y is 13. You find a point 10 units to the right of zero and 13 units above zero. Then you try more combinations. If x is 17, then y is 20. When you have several of these points located, you'll notice that they also form a straight line. Just by drawing that straight line, you can solve problems. If x is 35, and you've drawn your graph on "graph paper," with the little squares, you can simply count over to see what y is when x is 35. In a simple example like this, that may be more trouble than it's worth, but it illustrates one way a graph can assist you.

The two distances are called "coordinates." The first one is the "x coordinate," the second one is the "y coordinate." If you can remember that the x axis is the horizontal one, and x comes before y in the alphabet, perhaps you won't get confused as to which is which.

A line you graph may cross either the x axis or the y axis, or both. Where it crosses is called the "x intercept" and the "y intercept." Sometimes, knowing that spot will give you the answer to a particular problem.

Slope

When you begin plotting shapes to attack with geometry, you'll want to know what angles are involved. The concept of "angle" doesn't make much sense in algebra, however. It doesn't fit into our neat equations. Instead, we have a concept called the "slope," which is a way of describing the inclination of a line compared to the x axis and the y axis, and which takes the size of our units into consideration.

"Slope" is the slant of the line, its steepness. It's

described as the "rise" over the "run." That is, how does the "y coordinate" change for any change in the "x coordinate?" If y increases by 3, how much does x increase? In our first example, x = y, the slope is

$$\frac{1}{1}$$

That is, every time x increases by 1, so does y. "Rise" refers to the "y" coordinate and "run" refers to the x coordinate.

A slope of

$$\frac{3}{4}$$

means that every time the line moves 3 units "up," it also moves 4 units "to the right."

Slope looks like a fraction, and, in fact, can be used exactly like a fraction would be used in an algebra problem. That's another reason we use it instead of "angles." Later in trigonometry, you'll learn how angles are related to the slope.

If a line has a slope of 3/4 and the x coordinate moves 40, we can figure the y coordinate will move 30. We multiplied both numerator and denominator by the same number, just like we would have in an equation. The distinction between equations and graphs begins to blur, and it doesn't seem to bother us. We are becoming one of Them.

Curves

Repeating patterns can be translated into equations. Sometimes we can further translate these equations into graphs. When we do, we notice some bizarre things happening. Similar equations create visual designs that are similar to each other. Some kinds of equations always trace a straight line. Others always trace a curved line. Some create very specific curves, so common and predictable that the curves themselves have been given names. For example, if you throw a stone, it will follow a curved path known as a "parabola." And, if you graph a quadratic equation with two variables, that too will always create a parabola. We can use this little coincidence to calculate how to aim missiles and softballs. Or, we could examine a curve and learn something about the equation it represents. Many folks can look at the graph of a parabola and tell you if the quadratic equation that is its cousin contains positive or negative coefficients, for example.

Just as a quadratic equation often leads to two correct answers, so some equations, when graphed, lead to more than one line. You may see two curved lines, completely independent, perhaps looking like mirror images on a graph. Don't panic. You'll also see nice, wholesome lines with one section missing. Happens all the time.

When talking about curved lines, sometimes it's useful to create a new line to use as a reference. This new line is called the "asymptote." The main job of an "asymptote" is not to run into the curve.

I have to confess, I'm not a "visual person." Graphing still looks like cave drawings and magical symbols to me. Like finger painting or kissing, it's hard to explain the concepts (or the fun of it) without getting your hands wet or your lips puckered. So, don't despair that this sounds like a bunch of Mayan incantations. Once you're actually playing with them, the concepts will seem easier. Perhaps Dr. Jim can help make sense of it for us in the next couple of chapters. At the least, perhaps these new words won't seem so strange to you when you have to wrestle with them.

Dr. Jim Discusses Graphing

During the 16th century, "algebra-ists" busied themselves solving equations. They even staged contests to see who could do it better. Algebra was moving away from using words and toward increased use of symbols. In the meantime, the "geometers" were using their rulers and compasses to make constructions and draw complicated curves .

As you know, René Descartes (1596 - 1650), had the idea to associate the x's and y's of algebra with points in the plane, that is, on a piece of graph paper. In doing this, Descartes tied the two worlds — algebra and geometry — together for the first time. He created the rectangular coordinate system, a very powerful combination of tools that easily describes many facets of the world. For example, physicists can describe the motion of an object along a line (geometry) with an equation (algebra). Descartes' ideas were refined into the present day system called coordinate geometry or analytic geometry. It may be called graphing in your Real Algebra Book.

Imagine a piece of graph paper with two lines drawn like cross hairs in a siting scope. The vertical number line and the horizontal number line cross each other at their 0 points. Bull's eye is at the 0 point on each line. We call it the origin and label it (0,0). The first "0" tells you that you are located at 0 on the horizontal number line, while the second "0" tells you not to go up or down from there. As soon as this is done, any other point on the page is automatically assigned a number pair. For example, if the point moves to the right horizontally, the first number will change. After 1 inch (or unit), we will be at the point labeled (1,0). Another inch "eastward" (if this were a map) will put us at (2,0). If we now move upward (or north) on the paper for 1 unit we would be at (2,1). Read (2,1) to mean 2 inches to the right and then 1 inch up. Similarly (5,7) means to move 5 inches to the right and then 7 inches up.

It takes two numbers to describe each point. That's why a piece of paper is said to be two dimensional. On the number line (one dimension) we only need to know one number to locate ourselves.

Do you recall that wonderful toy that has a little gray screen and two red knobs, the Etch-a-Sketch ®? One knob controls the horizontal movement of the point and the other knob controls the vertical movement. This toy lets us play the same game Descartes played. He split the "Where is it?" question for a point into two parts: "How far left or right?" and "How far up or down?" Where is (-3, 4)? You just set horizontal knob at -3 and then set the vertical knob at 4.

You can locate any point if you know the horizontal number and the vertical number. It is customary to use the letter x to represent the horizontal number and y to represent the vertical number. We call the horizontal number line the "x axis" and the vertical one the "y axis."

In the point (3,-4), 3 is called the "x coordinate" and -4 is the "y coordinate." Your teacher and textbooks will soon talk about the point (x,y). To get to that point, you must move left or right x units on the horizontal number line and then move vertically y units. If you don't know what value x or y has, then you don't know where that point is. But as soon as you do, you know how to get there.

To describe a straight line on a piece of graph paper I could list all the points on the line using their coordinate pairs, but it would be a very long list. It's much easier to write the equation:

$$y = 2x + 3$$

If we use a Little Weenie Number as an x coordinate, we can figure out the y coordinate that belongs to it. For example, let's choose x = 3. Then the formula tells us that y will be:

$$y = 2 (3) + 3$$

which equals 6 + 3 or 9. The point has coordinates (3,9) so I move 3 units to the right and then up 9 units and put a dot there representing this point.

Now let's choose a different Little Weenie Number for x. Let's try x = 1. Then y = 2(1) + 3, which equals 5. The point is (1,5). Put a dot there too. This process is then repeated for several points, as we watch a visual pattern develop on the graph paper. You can visualize the straight line that I had in mind above. We have actually plotted some of the points. From the graph we can see where the rest belong.

Little Weenie Numbers are easy to substitute into the equation, and the resulting points are easy to plot.

I like to keep track of the x's and y's as I calculate

and plot points on an x-y table. So far it would look like
this:

$$
\begin{array}{c|c}
x & y \\
\hline
3 & 9 \\
1 & 5 \\
\end{array}
$$

My next Little Weenie x coordinate would be 0. I like to
use x = 0 often because the algebraic steps to find the y
coordinate are easy and it is easy to plot. It is on the y-
axis. This point is called the y-intercept of the graph of
the equation. The corresponding y value is 3. So now I
record the new point in the x-y table as (0,3). The table
now looks like this:

$$
\begin{array}{c|c}
x & y \\
\hline
3 & 9 \\
1 & 5 \\
0 & 3 \\
\end{array}
$$

Even more interesting is the graph paper with our 3 points plotted on it. As you can see, those 3 points line up in a straight line. It looks like this:

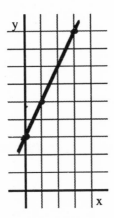

Equations that look like this:

$$y = 2x + 3$$

always turn out to be straight lines. Their general or standard form is:

$$y = mx + b$$

You can see why these equations are called linear equations. Linear means "in a line."

If some other Little Weenie Numbers stumble into your life looking for something to do, employ them as x coordinates and y coordinates. It keeps them occupied, and it's way cool to watch all the pints "line up."

Graphing Linear Equations: Slope-intercept Form

We can predict the graph of a straight line from its equation without doing lots of hard work. When we describe how one coordinate changes as the other does, the result looks like an equation. This relationship between coordinates can be stated in a couple of forms, which you'll come to recognize and love. You use the form to plot all the points on a graph having that relationship. When you connect those points, they'll form a line.

The "slope-intercept form" of the equation for a straight line tells you the slope and y-intercept of the line directly. It is usually written

$$y = m\text{x} + \text{b}$$

where m is the standard letter for slope and b represents the y-intercept. Remember "m" for slope, because it could stand for "mountain," and mountains certainly have some slope.

When you recognize that the equation contains the information you need, you can easily graph the line. The slope-intercept form is THE most useful form of a straight line equation, because it's easy to remember and to use.

Soon, you'll do what your teacher does: squint at the equation and just draw it. Here's how she does it. The equation

$$y = 2\text{x} + 3$$

has several things that a practiced eye will see. The "3" is the y-intercept, the place on the y axis where the line crosses. The "2" is the slope of the line. Slope is a way of measuring the steepness of a line. It is the ratio of the number of units you must move up or down for every x unit to the right. Here, $2 = +2/+1$, so this line moves +2 units (up) for each +1 unit it moves to the right. If you were drawing this line on the Etch-A-Sketch®, the "y knob" would turn twice as fast as the "x knob."

To draw the graph, you move up the y axis 3 units. From there you repeatedly follow the slope instructions; over 1 and up 2; over 1 and up 2, etc. Connect the dots and you've got the graph of the line.

When you see a linear equation in this form you can do a good job of guessing what the graph will look like even before you plot some points.

Consider

$$y = -3x + 7$$

Right away you'll recognize that this represents a straight line. You see that it crosses the y axis +7 units up. By rewriting the slope, -3, as -3/+1, you see that for each +1 unit to the right, the point moves down 3 units.

Starting from 7 units up the y axis at the point (0,7), move over +1 and then down 3 to (1, 4), move over +1 and then down 3 to (2, 1), move over +1 and then down 3 to (3, -2), move over +1 and then down 3, etc.

But now let's confirm these ideas by actually plotting points. Using Little Weenie Numbers, we make this x-y table:

x	y
1	4
0	7
2	1
3	-2

Sure enough, we are getting the same points that we already "guessed." Few things in life are as much fun as doing math problems that you thought were very hard AND getting them right.

The graph would look like this:

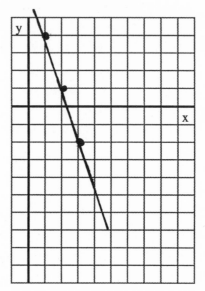

Textbooks try to be clever with linear equations. With some careful algebraic manipulation, all of the equations shown here can be made to look like the standard slope-intercept form shown above:

$$3x + y = 7 \qquad 6x + 2y = 14 \qquad 2y - 4x - 5 = 0$$

You want "y" on the left side. When you are asked to graph an equation of a line, the standard hint is to "solve for y" and use what you now know about the slope-intercept form.

Writing Linear Equations:
Point-Slope Form

The second most useful standard form is the point-slope form. You'll use it when you hunger to describe a line as an equation but are only given geometric information.

Your Real Algebra Book will show you this:

$$y - y_1 = m(x - x_1)$$

As before "m" represents the slope of the line, and (x_1, y_1) are the coordinates of a point on the line. The little "1" on the x_1 and the y_1 indicates that these letters are to be replaced by specific numbers. The plain unmarked variables, x and y, will remain as variables in the equation. It is a weird but standard notational practice in written math.

Perhaps you feel an irresistible urge to write an equation of the straight line with slope

$$\frac{2}{3}$$

that goes through the point (2,4). Because you know this standard form of the line you will just write:

$$y - 4 = \frac{2}{3}(x - 2)$$

Here the 2/3 represents the slope of the line, the "2" represents the x-coordinate of the given point, the "4" represents the y-coordinate. Notice the minus signs in front of the x and y coordinates of the point. The point slope form comes equipped with minus signs. Subtraction is part of the form, just as we had some subtraction in the Quadratic Formula. The dashes don't indicate that the coordinates you're injecting are negative numbers. If the point had been (-5, -7), the equation would have been written

$$y + 7 = \frac{2}{3}(x + 5)$$

Once you understand how this form is written, you can easily squint at it and figure out the graph. Move to the given point: 5 to the left and then 7 units down. From there, "do" a slope of

$$\frac{2}{3}$$

by repeatedly moving over 3 units to the right, then up 2 units. Now connect the starting point with the new points and draw a straight line.

Graphing Calculators

Instead of buying five or six copies of this book to send to each of your relatives for the next holiday season, you could, for the same money, purchase a graphing calculator. On a screen the size of a credit card, it will automatically make graphs of any kind of formula that you can dream up. Of course, we'd rather you bought five or six copies of this book.

To use a graphing calculator, simply type in equations, push the "Graph" button and watch as the calculator automatically chooses its very own Little Weenie Numbers, quietly twists its own internal Etch-A-Sketch® knobs, and draws the graph of the formula that was typed in. The gnomes who reside within these devices never get tired, although they get a little testy when their instructions are not clear.

These calculators are worth every dollar of their price. It will take several hours to figure out how to use one. But that's also a good investment. If it's any consolation, they are easier to figure out than your VCR. They will be useful in every math course you take. The Texas Instruments TI-81®, or the TI-82® are laid out well and perfectly adequate. If you expect to study calculus, you may want more bells and whistles. Talk with your future calculus instructor NOW and find out which kind to buy and learn to use.

By now you get the drift. Get a graphing calculator. Just don't let your little brother know that you have one. You'll never see it again.

Sets

A "set" is a group of things that can be described in terms of their common characteristics. The word "dog" refers to millions of individual slobbering animals. Each of them share enough characteristics to be included in the "set" of "dogs." Simple drooling isn't enough, for example. If it was, then half the noble college students who attend spring break on the beach would have to be considered "dogs." The set of dogs is much more elite than that. There's the "four paw" requirement, for example, plus fur, and a minimum standard of common sense. Still, it's a big set. Other sets are more restrictive. The set of "algebra books" is smaller than the set of dogs. The set of "interesting algebra books" is smaller still.

Obviously, there's a certain amount of overlap. Every "interesting algebra book" is also a member of the larger set "algebra books." There are, however, very few dogs that would also be included in that set, nor very many books of any kind that can meet the requirements of the "dog" set. But every "dog" is also an "animal." So, definitions restrict sets, but one set can include another or a portion of another.

More restrictive definitions usually result in smaller sets. When you completely define "planet Earth," your set will only contain one member, for example. When you define the "integers between 3 and 5," the set will consist of the number 4. It is even possible to define sets with no members. "All the numbers greater than 3 that are also smaller than 3" defines a Zen-like bunch of numbers that simply don't exist. This is called the "empty set." No members. Over time, and with many government grants, mathematicians have developed rules for manipulating sets with no members and created new

220

courses that must be taught.

But didn't I originally describe a set as a "group" of things? How can one item, or no items, be a group? Well, of course, it can't. This is a situation when trying to be scrupulously accurate with your words gets in the way of saying what you mean.

It's easy to understand sets if you think of them as groups of things that share specific characteristics, such as "all the numbers greater than 30." A tiny mental step allows us to admit that, for our purposes, a "group" may contain only one member or even no members. The set of "black-footed ferrets" has declined in recent years as the population has dwindled toward extinction. Should they continue to die off, at some point the set of black-footed ferrets will contain only one member. If the last ferret dies, the set will be empty, even though the definition will survive.

And perhaps someday the sets of "algebra books" and "interesting algebra books" will each include all the members of the other.

Dr. Jim Explains How Mathematicians Eat Ice Cream

Tradition demands that math books must all be written in a certain writing style. Until you learn to translate that style, it will frighten and confuse you. Trust me: mathematicians are nice, fun-loving, madcap wags. Some of my best friends are mathematicians. They are rarely dangerous. You just have to get used to the way they talk:

Let ice cream equal (ic), and spoon equal (s). It is obvious by previous sesquipedalians that for any bowl (b) of (ic), 1 (s) is sufficient to consume 1 (b) if that bowl is within the subset (b)< q, where q = 1 quart. From this one can easily see that E = f(b) where E = eating (ic) and that E is also directly proportional to hunger (H), and exponentially proportional to the surface-to-volume ratio of the subject mathematician (M) defined as (g).

Immediately, several things become obvious. As (H) and (g) increase, one must resort to literary hyperbole to graph the related functions, and perspicuity sublimates to a miasmic obfuscation, wherein sagacity is transmogrified into an oblique hubris of the lexicon. Therefore, we can see that E=(b)(ic) - (s) at a rate defined by g(H), according to the Gilligan/Belushi Principle, which will be discussed in greater detail in a later chapter.

It should be noted that (E) is a non-invertible process, as can be seen by

$$M(E) \;=\; b$$
$$E(M) \;\neq\; b$$

Attempts to perform the inverse of (E) should be avoided by the student, as they can be very messy. Although the advantages of this method are clear, it becomes even more powerful in conjunction with other processes. This should become obvious in the following example: Let (PG) equal a beautiful woman of approximately equivalent age to the subject mathematician, and previously described as "pretty girl." Let f(k) equal kissing, where f(k) = Mk(PG) and m^2 equal moonlight. Therefore:

$$P = M + (PG) + (E) \text{ ic } + m^2 + f(k)$$

where P = pleasure. The student should memorize each of these steps, although the processes described are beyond the scope of this text. You'll get no ice cream in this class. It should be noted that, without calculus, (P) is undefined and irrelevant.

Roots

When you solve a quadratic equation, you'll get no more than two correct answers. One of these might be zero. One might be 23 and the other (-23). You'll have to choose which answer makes sense in the real world. When you solve a cubic equation, or third-degree equation, (something's raised to the third power, but nothing's raised to a higher power) you'll get as many as three correct answers. A fourth degree equation can yield as many as four answers.

Usually, these equations are presented as expressions equal to zero. Any number you can substitute for x that makes the whole expression equal to zero is called a root.

You try to manipulate a quadratic equation until it becomes a multiplication problem whose answer is zero, like this:

$$(x + 4)(x - 5) = 0$$

Because the only way a multiplication problem results in zero is if one of the factors is zero, either (x + 4) equals zero or (x -5) equals zero. If (x + 4) equals zero, then x must equal (-4). If (x - 5) equals zero, then x must equal 5. Both negative 4 and positive 5 are "roots" of this equation.

A "root" may satisfy an equation but make no sense in the context of the real world problem. So we don't want to say "x equals negative 4." We don't quite know that. What we know is that x equals either (-4) or 5. The equation works just as well with either one. We have reduced our correct, rock-solid, dead-on, bet-the-farm answers down to two possibilities. Those two are each roots. They each make the equation equal to zero.

You'll also hear the term "square root" and "cube root," which are a little different. The "square root" of a number is the number you multiply by itself to get your original number. Three times three equals nine. Another way of saying that is "3 squared equals 9." And the square root of 9 is 3. It's probably called a square "root" because it is the root of the equation

$$x^2 - 9 = 0$$

Cube roots extend that process:

$$3 \bullet 3 \bullet 3 = 27$$

and

$$\sqrt[3]{27} = 3$$

Because a negative number times itself equals a positive number, the square root of 4 can be either 2 or negative 2. We have to choose which one makes sense. Before all the math teachers write angry letters, let me apologize and correct myself. It's not quite that simple. Although a negative 2 squared equals positive 4, they won't let you say that the square root of 4 can be either 2 or negative 2. The symbol

$$\sqrt{4}$$

will always represent only positive 2, even though we both know that 4 has two square roots. If you want to use symbols that to show negative 2 is also a square root of 4, you have to write

$$-\sqrt{4} = -2$$

A root is a number that makes some expression equal to zero but which may not be the only number that does. When you make a graph of an equation, each root identifies where the line crosses the horizontal axis.

Inequalities

Our scale won't balance with different weights on each side. Despite this inconvenience, often one item is larger than the other. Tom sinks our rowboat and Mary doesn't. One sack of beans weighs more than the other. You get more stuff with your paycheck than I do. Sometimes, this inequality is all we know. Without additional information, equations won't help us much. Tom equals x and Mary equals y. A thousand manipulations fail to increase our insight. On the other hand, sometimes simply understanding the inequality is enough. Shall we row across the lake with Tom or with Mary? Which sack of beans should I carry in my backpack, and which shall I let you carry? Sometimes we can express an unknown as part of an "inequality" and narrow our choice of solutions to a useful size. We may not be able to say exactly how much Mary weighs. But perhaps we'll learn if our additional weight would sink her boat. And truthfully, that's all we really care about anyway. Inequality is expressed with this symbol:

$$>$$

It can face either left or right. The larger item is

on the open side of the symbol, the smaller item on the pointy side. So, if you see:

$$x > y$$

you understand that "x is greater than y." If you see:

(Your brain) < (My brain)

you will understand that you have just been both lied to and insulted. We pronounce the inequality sign either "is greater than" or "is less than" depending on which way it faces. "Greater than" and "less than" are nice, friendly, real-world concepts that don't bother us much. But sometimes they feel threatening, like when they involve negative numbers. Which is greater, negative 10, or negative 20? The short answer is: negative 10. On a number line, a number is greater than any number to its left.

When you fill a cup with water blindfolded, at some point it will probably overflow. Before it spills, it either contains exactly 1 cup of liquid or less than a cup. Until it spills, the amount of water in the cup is "less than or equal to" 1 cup of liquid. This condition is described by putting an inequality sign over half an equal sign. "Less than or equal to" looks like this:

$$\leq$$

The concept of "greater than or equal to" is expressed by reversing the direction. The small end of the symbol always points at the smaller item.

If two items do not equal each other, they put a slash through the equal sign, like this:

$$\neq$$

and pronounce it "does not equal."

So, "cat does not equal dog" would be written "cat $\neq$ dog." "Watermelon > grape" is generally a true statement. "My income ≤ my desired income" is generally true; that is, some people feel they earn enough, many wish they earned more. People rarely complain that they're earning more money than they'd prefer.

Inequalities express limits. They define the boundaries of whatever little scene we're trying to describe. If the bag of sugar weighs less than a hundred pounds, we know it doesn't weigh a million pounds. If x does not equal 43, then we've reduced our guesses by one choice.

A series of several inequalities may define the limits even further. If we're looking for a number greater than three, but less than ten, and it doesn't equal seven, we've got only five integers to choose from. In the real world, if you don't have enough information to create an "equation," perhaps an "inequality" will do you some good. If you can express a series of inequalities, you may be able to reduce your choices.

Inequalities can be manipulated with some of our old tricks. If Tom's side of the teeter-totter is drooping toward the ground, and Mary's is high in the air, you can add a box of apples to both sides without causing either one to move. Tom's side is still heavier. In fact, you could add a Chrysler to each side, and Tom's would still be heavier. You can add the same amount to each side of an inequality without changing the inequality.

You could also subtract each of their identical cameras. No effect. You can subtract the same amount from each side of an equality without changing the inequality.

You could try the experiment 40 times in the same day, and the result would always be the same. Tom times 40 on one side, Mary times 40 on the other side, and Tom's side is always heavier. You can multiply both sides of an equality by the same amount without changing the inequality.

Without a grisly and tasteless analogy, it's a little hard to illustrate division. Perhaps we need to replace Tom and Mary with an apple and a grape. The apple > grape. If we cut them both into three pieces, the apple side is still heavier. Which is to say, you can divide both sides of an inequality by the same amount without changing the inequality.

The only exception involves multiplying and dividing by negative numbers, which you have probably already felt an urge to try. Little Weenie Numbers illustrate how this works. Three is greater than two. If we multiply them both by negative 1, we get negative 3 and negative 2. But negative 2 is greater than negative 3. This is the opposite of our original relationship. You may spend a Saturday afternoon creating inequalities and then multiplying them by negative numbers. Or, you may choose to accept the wisdom of all the Greeks and Babylonians who spent their Saturday afternoons merrily calculating. They discovered this: When you multiply both sides of an inequality by a negative number, the inequality reverses. Should you somehow multiply both Tom and Mary by the same negative number, Mary will become heavier than Tom. It would be like turning yourself upside down. In your new frame of reference, Mary's side is lower and Tom's side is higher. Upside-down Mary now looks heavier than upside-down Tom.

Similarly, if you divide both sides of an inequality by a negative number, you must also reverse the inequality sign. Assault this concept with your own arsenal of Little Weenie Numbers until your brain surrenders to it.

Another strategy for manipulating inequalities is cross multiplying. Most of the time, if you're careful not to move the numerators across the inequality sign, the inequality won't be affected. If:

$$\frac{x}{y} > \frac{a}{b}$$

then

$$xb > ay$$

Notice that both x and *a* (the numerators) stayed on the sides of the inequality sign they started on.

One big exception. If one of the denominators is a negative number, the inequality reverses when you cross multiply. "Less than" becomes "greater than" and vice versa. Try it with small numbers. On the other hand, if both denominators are negative, the inequality does not reverse. This gives even more weight to Dr. Jim's position that cross multiplying is a dangerous trick in the hands of guys like you and me.

Dr. Jim Describes Functions

We often associate one number (called the output) with another number (called the input).
Some examples include:
•   Tell me the weight of the letter (input), and I'll tell you the postage (output).
•   Tell me the number of miles per hour (input) that you were over the speed limit (and the state where you were driving), and I will tell you the amount of the fine (output).
•   Tell me the amount (input) in the bank account (and the interest rate) and I will tell you the amount of

interest (output) you will earn in a year.

• Tell me how long you have been driving at 55 mph (input) and I will tell you how far you have gone (output).

Each of these relationships is an example of a function. In algebra class you'll learn about many different functions.

My favorite image of a function is a stainless steel machine that my grandma had on her back porch. It separated the milk from the cream in fresh cow's milk. You poured the milk into the top. Then you turned the crank and the parts started whirring. If you were lucky, you had remembered to put a bucket under each of the spouts, one where the skim milk came out and the other where the cream came out. You poured the cow's milk into the machine, and it produced the cream.

Sometimes we describe functions as equations. If we want to reuse the equation to use in similar future situations, we abbreviate the roles each number and symbol plays in it. We call these reusable equations "formulas."

Any formula can be thought of as a function if it has inputs and outputs and each input always generates one specific output. When you gas up with 6 gallons and the price of gas is $1.15 per gallon, the pump figures out the total price:

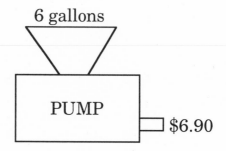

Mathematically, we'd write the PUMP function this way:

PUMP(6 gallons) = Total price or P(6) = $6.90

Its formula is written

$$P(x) = 1.15x.$$

Formulas, and Why Dr. Jim Loves Them

A formula is a mathematical recipe, a standardized, reusable set of instructions. You plug the information from your situation into the formula to create your desired results.

When I make brownies, the box tells me to use 1 egg and 3 tablespoons of water, and then stir exactly 50 times. It gives me the exact relationship between my raw materials and also tells me what to do with them. Today I want to make a triple batch so I'll use 3 eggs and 9 tablespoons. But do I now stir 150 times?

No. I subject any amount of raw materials to the same procedures.

A formula is like a little number-eating machine that eats numbers and, using some weird recipe, spits out another number. Here's is a record of several tries that I made with this toy.

IN	1	2	3	4	5	6	7
OUT	1	3	5	7	9	??	??

You could probably figure out what the last entry ought to be. If I feed the machine a 7, it should spit out a 13. But what is the recipe? You notice the relationship between the first IN and OUT numbers. There are several possibilities. Then you look at the second pair to see if they shared any of those patterns. Pretty soon, you devise your own formulas. For our example, we might have come up with this formula:

Take the IN number, and divide it in half. Now add 1. Multiply this new number by 4 and set aside. Prepare a negative 3. Combine the -3 with the number set aside earlier. Garnish the OUT with some parsley and serve.

You probably figured out a simpler recipe. Perhaps, "double the IN number and subtract 1."

Formulas like this can help us understand why some of the things that seem weird actually help the game remain consistent. For example, why does $2^0 = 1$? Let's look at a pattern that involves exponents.

IN	1	2	3	4	5	6	7
OUT	2	4	8	16	32	64	

The IN numbers increase by ADDING 1 each time and the OUT numbers increase by MULTIPLYING by 2 each time. If you went from right to left, the IN numbers decrease by subtracting 1 and the OUT numbers are cut in half each time. No problem.

Let's extend this table to the left:

IN	-1	0	1	2	3	4
OUT	?	?	2	4	8	16

Mathematicians want consistency, even when it doesn't make much actual sense. So under the 0, they would put 1 (That is half of the 2) and under the -1, they would put 1/2. This pattern can be also be described using exponents in a formula $2^{IN} = OUT$. That is, $2^2 = 4$, $2^5 = 32$ and $2^0 = 1$.

In order for the game to be predictable, they really had no choice. This is how exponents need to work.

Functions

The Lunch Lady has a system for assembling school lunches. She takes a plate, adds a peanut butter sandwich, slops on a gob of Jell-O, adds 2 cookies, then a banana. Every plate is an empty canvas for her artistry. Each time she completes her routine, she gets an identical lunch.

When she teaches a new assistant her routine, she begins by carefully explaining each step. Once the assistant understands the fine points of the procedure (like, "don't drop the plate") she can express the entire routine in much fewer words. She can say "make lunch."

In algebra, a "function" is a procedure we abbreviate like that. It may be simple, like "double the number" or it may be complicated, like "square a number and then subtract 4 times the number."

If you start with a number and perform a function on it you'll end up with exactly one answer. This is different than, say, seeking roots, which can give us a pocketful of answers that work in the equation, some of which may be absurd.

If a procedure is a function, and you apply it to a number, you'll get one answer, but not more than one. You start with ten plates, subject them to your procedure, you wind up with ten meals. One meal per plate.

There's a specific relationship between the plate and the resulting meal. If the "plate" is a changing quantity, like time or motion, the meal will change right along with it in an exact way.

For example, John always sneezes one hour after Mary does. Mary sneezed at two. When will John sneeze?

We can describe this as a function. A little stylized f means "function." We're going to make the time of John's sneeze "a function of Mary's:"

$$f(M) = J$$

This is read as "function of M equals J." The parentheses no longer mean multiplication. When you see the little f you know you're dealing with a function. The procedure that the function uses for its calculation will be given to you in symbols rather that saying "John always sneezes 1 hour after Mary." In this case, you'd see:

$$f(M) = M + 1$$

The letter M has a couple of important jobs here. It represents the time that Mary sneezes. And it is also used to describe how the function, f, operates.

The information given in the example is that Mary sneezed at two. To find out when John sneezed you would follow the "formula" given for the function f. In other words, what they want you to do is substitute "2" into the "M + 1" and complete the arithmetic. If this were your

Real Algebra Book, you'd be asked to "evaluate" $f(2)$. Your work would look like this:

$$f(2) = 2 + 1 = 3$$

Because you have substituted the "2" in place of M on the left side, you would also substitute it for M on the right side. 2 plus 1 is 3 o'clock, which is when we'll want to move away from John.

A function is any formula or procedure that transforms one number into another. The set of numbers that are fed into one of these procedures is called the domain. They almost always represent these about-to-be-transformed numbers by the letter x. For the Lunch Lady, the domain is the stack of clean plates. The output numbers are usually represented by y. When collected together, these "y" numbers form a set called the range of the function. She knows this set as the meals that were served.

Once defined, the symbol f is shorthand for the entire procedure. First they define the process, then they abbreviate it. Whenever you see the $f(\ )$, they want you to perform the f-process on whatever's inside the parentheses. If your function adds 10, when you perform that function on the number 8 you'll get the number 18.

When you see the symbols

$$y = f(x)$$

you'll know that they have taken some poor unsuspecting generic x and fed it into the function named f. When it came out it got stamped as a y. It is automatically understood that the x came from the domain of f and that y will be a member of the range of f. When functions are in the vicinity, all the numbers know whether they are in the domain or the range.

It is too bad that math books don't write it as

$$(x)f = y$$

Reading from left to right, it would mimic the actual process:

"x passes through f and is transformed into y."

It's a great idea! Unfortunately, it is only shown to mathematics students in graduate school. By then it confuses them because it makes so much sense.

Our lunch function might be described like this:

$$f(\text{plates}) = \text{meals}$$

And the definition of the function would be:

$$f(\text{plates}) = \text{plates} + \text{peanut butter} + \text{Jell-O} + 2 \text{ cookies} + \text{banana}$$

If we have 43 plates, we might wonder how many total lunches we'd get if we apply our plates function.

$$f(43) = ????$$

After a lot of hard calculating, we discover that we have 43 plates full of food after we're done. The meals are a function of how many plates we started with.

If we start with 43 empty plates, we'll always end up with 43 meals.

The letters "f", "g," and "h" are commonly used for functions. If a problem has more than one function, odds are good your Real Algebra Book will use these letters in either upper or lower case to represent them. And, if you see these letters in a problem, (especially in italics) they are likely to be describing functions. But, sometimes, a writer will fool you and use the first letter of the activity. Time might be represented by "T," distance by "D," for example.

If the authors of your Real Algebra Book wanted to give a name to the procedure that triples things, they would write $f(x) = 3x$. You would know that "f" stands for tripling things. f(scoop of chocolate chip) is a tall ice cream cone. If they wanted to describe the process of squaring a number and then subtracting 4 times the number, they might name it g and write: $g(x) = x^2 - 4x$. The left side of the equation names the process. The rest tells what it does and in what order. The letter x represents all the Little Weenie Numbers that could be plugged into the formula. As you read the symbols, think about what is happening to those unfortunate individuals. It would be more clear if they wrote these functions like this:

$$f(\_) = 3 \bullet (\_) \quad \text{and} \quad g(\_) = (\_)^2 - 4 \bullet (\_).$$

Algebra is much more powerful than arithmetic because it uses letters to represent unknown numbers. A similar gain in power occurs when formulas and procedures are abstracted and named with letters. Just as x changes meaning from problem to problem, the functions represented by the letter f will change from problem to problem. Don't worry. It only gets better from here on.

Bunnies 'Do' Functions

Easter Bunnies hide colored eggs. Some kids in Cleveland figured out that the bunny who handles their district hides eggs in a predictable way. The night before Easter, they leave carrot sticks out for the cuddly rodent.

This particular Bunny always eats the carrot stick, then hops 10 feet north and hides an egg. Being bright, mathematically inclined kids, these 6-year-olds decided they could reduce this pattern to a function and employ algebra to improve their egg-locating efficiency. It's what you or I would have done at their ages, too. They realize that the location of the egg is a function of the carrot stick. They write this with their crayons:

$$f(\text{carrot}) = \text{egg}$$

Which they read as "function of carrot is egg" or "egg is a function of carrot." They describe the function itself this way:

$$f(\text{carrot}) = \text{Go 10 feet north of carrot}$$

We know the relationship is a function, and not something else, because every carrot stick location results in only one egg location.

The kids decided to put carrot sticks in the center of a sidewalk, one on each crack. Sure enough, on Easter morning they found a neat line of eggs 10 feet north of the center of that sidewalk. Wherever the sidewalk curved, the line of eggs curved in exactly the same shape.

The location of each carrot was part of the domain. The distance to each egg was in the range of the function of f.

When the kids placed kite string along the ground from egg to egg, they said they were graphing the function. The grown ups told them to cut that out, they were wasting string. Couldn't they just watch cartoons, like all the other kids?

In Detroit, the designated bunny operated on slightly different instructions. He always hid two eggs, one 10 feet north of the carrot, and one 10 feet south.

This variation was probably the result of a clause negotiated by union representatives and the Easter management team. It was still pretty easy to find the eggs. There was still a firm relationship between carrots and eggs. But the relationship was no longer a function. Now, any carrot yielded two different egg locations. In order to be a function, it could only yield one. But of course, because of the contract, they can't fire the bunny.

In Princeton, N. J., the bunny takes delight in confounding the children. After corresponding by e-mail with their buddies in Cleveland, the kids decided to lay the carrot sticks three feet apart along the straight sidewalk. They wrote the word "origin" on the sidewalk. Three feet east of that and every three feet along the sidewalk, they put a carrot stick.

They found their first egg exactly nine feet north of the first stick. But the second egg was not nine feet north of the second carrot stick. It was 36 feet north. The kids were puzzled for a moment. Then they realized the silly rabbit was dropping eggs the square of the distance the carrot was from the bowl. The second stick was six feet from the bowl. 6 times 6 is 36. The third carrot was nine feet from the bowl. The egg was 81 feet north of that.

The kids got more exercise than eggs. The bunny, who was great at math, but not very smart, didn't have time to finish his territory. The kids noticed right away that if they connected the eggs with kite string, the string would trace a specific kind of curve, called a parabola.

The Los Angeles territory is handled by two bunnies, Flopsy and Mopsy. Flopsy always hops 20 feet north of the carrot stick, then south half the distance the carrot is from the bowl. So, if the first carrot is 4 feet east of the bowl, he hops 20 feet north, then comes back south two feet and hides an egg.

Mopsy follows Flopsy. She hides an egg three feet South of Flopsy's egg.

The kids in LA recognize this as two separate functions. There's the Flopsy function (F), which is:

F(carrot) =   20 feet N of carrot, minus one-half the carrot-to-bowl distance

Mopsy's egg is a function of Flopsy's. We'll use the letter M to represent the Mopsy function:

M(F(carrot)) = 3 feet South of F(carrot).

If a kid only wants Mopsy eggs (they're chocolate), he'll find the location of any carrot stick and go through the two functions.

If you connect the results of different functions with kite string, you wind up with different shapes on your lawn. Some functions trace straight lines. Some trace curves of various kinds. Finding the points and connecting the dots is called graphing the function.

The Easter Bunnies start by knowing their preferred function. They find a carrot stick, go through the procedure, and hide an egg.

The kids start by finding a few eggs and trying to figure out what in the world the rabbit is up to.

Your algebra book might take either approach. It may begin describing functions as two quantities that are related in a specific way. It will say we found this carrot stick here, and this egg there, and then this carrot stick, and this egg. You will look for the procedure that created the pattern. Or, it might describe functions as an abbreviation for an activity and you'll have to hide the eggs yourself.

You can tell a lot about the original function by

looking at the shape the kite string assumes on the grass. You can tell if it really is a function. You can tell if your function involves positive numbers, negative numbers, or both. You can tell if it's got an x squared in it or not. You can also solve problems by graphing that might be extremely difficult to solve with an equation.

And, of course, you'll save a lot of time on Easter morning.

Solving Systems of Linear Equations

The game takes a new turn now. There will be two linear equations that each have two variables. The job is to figure out what single pair of numbers will work in both equations. Your book may call this a "simultaneous solution" because the same pair of numbers must work in both equations. Two different but related equations are called a "system."

Perhaps you can write the process of making pizza as an equation. A different equation might involve the pizza-delivery-person's driving habits. Clearly they are different equations. But they both affect when your pizza arrives.

If we graph a linear equation, we get a straight line. When we graph a system of two linear equations, we find out where two straight lines cross. There are two ways to solve systems of linear equations. Let's invent two equations to use as guinea pigs.

First, we throw a dart at our graph. It sticks at a point four inches right of zero, and two inches below the x axis. That is, it sticks at point (4,-2). We decide that's where our two lines will intersect. Grabbing the first Little Weenie Number that walks by (happens to be five), we

241

decide that one line will move five inches plus an unknown amount vertically every time it moves one inch horizontally.

$$y = 5x + ?$$

When we plug in (4,-2), we get

$$-2 = 5(4) + ? \quad \text{or} \quad -2 = 20 + ?$$

Subtracting 20 from each side, we get

$$-22 = ?$$

So the first equation will be
$$y = 5x - 22$$
For the second equation, let's use the form

$$3x - 4y = ?$$

Again plugging in (4,-2) we get

$$3(4) - 4(-2) = ?$$

So the "?" is 12 plus 8 or 20. Our second equation will be
$$3x - 4y = 20$$

In your Real Algebra Book, the exercise would look like this:

Solve: $y = 5x - 22$
$3x - 4y = 20$

It's easier to solve this "system" than it looks. There are two main ways to do it: the substitution method and the elimination method. In the next chapter we'll apply the substitution method to this problem. If you feel

242

a little panic, remember this: you already know the answer. The solution is the ordered pair (4,-2). We just created the problem together, remember?

After that, we'll apply the elimination method to a similar problem.

================================

The Substitution Method

It was midway through the third quarter on a blustery November Saturday. The gray sky, cold wind, and high humidity had taken their toll on the gridiron. Our team was determined to break the 0-0 tie. We looked at our opponent, a nasty, ugly, brutish monster that looked like this:

$$\text{Solve:} \quad y = 5x - 22$$
$$3x - 4y = 20$$

"Huddle up here. What can we do? We know how to solve equations that have ONE letter in them, not two. All right! Let's make it happen right now! We're going to change this pair of equations into one equation! "Suzy, you be a wide-out to the right. Joey you line up directly across from the "y" in the second equation."

$$y = \ 5x - 22$$
$$\underline{\qquad 3x - 4y = 20 \qquad}$$
$$\text{J} \qquad\qquad \text{S}$$

"When I say 'Hutt-Hutt,' Suzy you run wide around the right side of the top equation and begin blocking the 5x - 22 back down the page. Joey, you . . . HEY Joey, pay attention. Look at the bottom equation. See that big "4y"? You've got to blast the "y" and open a big hole there. But you've got to leave the "4" just standing there wondering what happened. Because that's where Suzy will push the (5x - 22). OK? Everybody ready? Let's get 'em!"

"Can we do that? I mean, is it legal?"

"You bet! Our first equation says what 'y' equals. We can always replace something with an equal item. If 'y' equals 'orangutan,' then we can substitute 'orangutan' anytime we see 'y'. Lucky for us, that first equation didn't have any orangutans. Just some Tiny, Puny Numbers that will be very sorry they suited up for the game.

"Hutt-Hutt!"

$$3x - 4(5x - 22) = 20$$

"Way to go everybody! Now we have one equation with one unknown. We can win this game!"

Here's how they finished their game. Multiplying out, they got

$$3x - 20x + 88 = 20$$

After combining like terms and subtracting 88 from both sides, they got

$$-17x = -68$$

Dividing both sides by -17 yielded x = 4

Pretending that 4 is a Little Weenie x-Number and plugging it in for the x in the first equation, they got

$$y = 5(4) - 22$$

So, y = -2. The solution to this system of equations came out just as planned, (4,-2).

During the post-game analysis, the local radio station interviewed Joey.

"Yes, sir. That was one cool play. We knew that sometime during the game, we'd get to use our Substitution method. It has worked well for us all season. And when it works, it's just a great feeling. It was great to be part of a team that was able to take two equations that had two variables and IN ONE PLAY make them into a single equation with only one variable. I mean it just doesn't get any better than this! Can I say hi to my Mom?"

The goal is to manipulate the system of two equations into one equation with a single variable. The substitution method does that by taking an expression that is equal to "y" in one equation and using it to replace the "y" in the other equation. In this exercise, we used "y" because one equation was already solved for "y." The choice of the variable doesn't matter. Choose the one that is easier to solve for.

It's really common sense. If you know that x equals 10, you'd automatically substitute 10 for all the x's in a problem. If you know that x equals ($y^2 + 43y +15$) your substitution decision may be less automatic, but just as reasonable. As Joey said, it is one cool play!

Elimination Method

The Elimination method is based on the principle that adding "equals" to "equals" gives "equals." That is, we can add the same thing to each side of an equation. If $x = 10$, we can add x to one side of the equation, and 10 to the other. The remarkable conclusion of this line of thinking is that we can add one true equation to another true

245

equation, and the sum will also be a true equation. We can add "x=10" to another equation (adding the x to one side, the 10 to the other) and the sum will be a true equation. But why bother?

The goal is to arrange things so that when we add the two equations, one variable or the other gets "eliminated." Here is the next problem in your Real Algebra Book:

Solve: $y = 3x + 11$
         $4x + 2y = 2$

Begin by lining up the equal signs and arranging the equations so that like terms are in the same columns. In some exercises this will be very easy to do. Sometimes it helps to draw a line like we're going to add up a bunch of numbers.

$$\begin{array}{r} y = 3x + 11 \\ \underline{4x + 2y = \qquad 2} \end{array}$$

The plan is to add the left-hand sides of each equation and set their sum equal to the sum of the right-hand sides. Notice that if we did it now, we'd end up with one equation, but it would have both x's and y's in it.

$$\begin{array}{r} y = 3x + 11 \\ \underline{4x + 2y = \qquad 2} \\ 4x + 3y = 3x + 13 \end{array}$$

This is no help. Backup.

$$\begin{array}{r} y = 3x + 11 \\ \underline{4x + 2y = \qquad 2} \end{array}$$

The goal is to eliminate one variable. Some clever person had a great idea. "If that top equation just had a (-2) in front of the 'y,' then the -2 and the +2 would add up to zero." We can multiply the left side of the equation by -2, as long as we do the same thing to the right side:

246

$$(-2)(y) = (-2)(3x + 11)$$
$$\underline{4x + 2y \quad = \qquad\qquad 2}$$

The new top equation now looks like this:
$$-2y = -6x - 22$$
$$\underline{4x + 2y = \qquad 2}$$

Adding them up again,
$$-2y = -6x - 22$$
$$\underline{4x + 2y = \qquad 2}$$
$$4x + 0y = -6x - 20$$

or more simply:

$$4x = -6x - 20$$

This is what we wanted: One equation with one unknown. Adding 6x to each side, we get

$$10x = -20$$

Dividing both sides by 10, we get

$$x = -2$$

Remember that we are not looking just for x. The answer to this type of exercise is an ordered pair of numbers (x,y). Plug -2 in for the x in either equation. The top equation looks easiest:

$$y = 3(-2) + 11 = -6 + 11 = 5$$

The solution to this system of equations is (-2,5). To check your calculations, substitute -2 for x and 5 for y in BOTH of the original equations. It would look like this:

$$y = 3x + 11 \qquad 4x + 2y = 2$$
$$5 = 3(-2) + 11 \qquad 4(-2) + 2(5) = 2$$
$$5 = 5 \qquad\qquad 2 = 2$$

The graph would show two lines that intersect at the point (-2,5).

Two Lines Cross

We have seen how algebra solves this problem.

$$y = 5x - 22$$
$$3x - 4y = 20$$

Here's how the graphing work would go. The first equation is in slope-intercept form. Using the "squint method" and what we know about that form, we see that the y-intercept is -22 and the slope is 5. Using Little Weenie Numbers, we also can get an x-y table:

x	y
0	-22
2	-12
4	-2
3	-7
5	3

The graph would look like this:

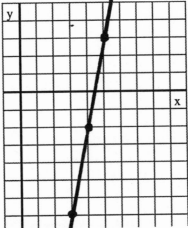

To graph the other equation: 3x - 4y = 20, begin by using zero, the easiest Little Weenie Number, both for x and for y. It's a handy trick sometimes.

x	y
0	-5
$\frac{20}{3} = 6\frac{2}{3}$	0
2	yuck
4	-2

Next choice for x was 2. It looked like it would be easy to find the y value. It wasn't. So we skip it. Try another x. If x is any odd number, then the y value will contain a fraction. So skip them. x = 4 works nicely.

Here's the graph with the new line:

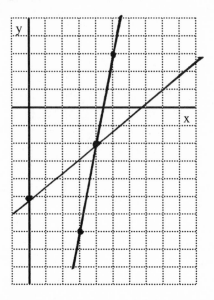

They intersect at (4,-2). Your graphing calculator will show you the same picture after you get the window dimensions adjusted.

Simplifying Radical Expressions

A radical expression is algebraic expression that contains one or more radical signs. It may include square roots, or cube roots, etc.

There are several reasons why you won't find anything in this book about this topic. For one thing, we've written this book for beginners. You can go a long way in algebra before you need to know about these monsters. But the main reason is that it is very hard to come up with important mathematical examples where knowing how to simplify a radical would be important. There are far more interesting and useful things you should learn about before becoming expert on this topic. In the mathematical symphony, it is a minor movement and often omitted in performance.

For similar reasons, we have glossed over material on manipulating algebraic fractions.

If your algebra teacher says she disagrees with these statements, find out about her reasons. At least, it will probably be an interesting discussion.

Other Algebras, by Dr. Jim

There are several algebras: college algebra, modern or abstract algebra, and Boolean algebra. The Islamic scholar al-Khwârizmî wrote the "al-jabr" treatise around 820 to explain some standard methods of solving equations. The English word, "algorithm" is derived from his name. An algorithm is an often repeated sequence of steps. Millions of children are learning the pencil and paper "algorithms" to add, subtract, multiply, and divide whole numbers as you read this.

College algebra is the name of a one-semester course offered at many colleges and universities that immediately precedes trigonometry and calculus. Typically, it consists of a hodgepodge of topics that will be needed in calculus. It includes all of the topics mentioned in this book together with more advanced discussions of functions, the graphing of many sorts of equations, introduction to sequences and series, and solutions to systems of linear equations using matrices and determinants. When these topics are combined with trigonometry, the course is usually called "pre-calculus." Courses like beginning and intermediate algebra are prerequisites to college algebra. In high school, most of these topics are included courses called Algebra I and Algebra II.

Modern algebra (or abstract algebra) is a branch of mathematics that is usually first introduced in a junior-level college course for mathematics majors where several cool topics are explored. Here's a taste of one, called clock algebra.

Suppose that instead of working with the real number system, you did all of your counting and arithmetic on an old, round, 12 hour clock. Then the basic number facts would change. For example:

$$3 + 7 = 10$$
$$5 + 8 = 1$$
$$11 + 10 = 9$$
$$x + 12 = x$$

Notice that 12 plays the role of zero; anything + 12 = anything. In fact, let's change the clock face so that 0 is at the top instead of 12. Then anything + 0 = anything, just like always.

These sample multiplication facts are less familiar:

$$2 \cdot 4 = 8$$
$$2 \cdot 8 = 4$$
$$5 \cdot 6 = 6$$
$$4 \cdot 3 = 0$$

This is interesting! Our trusty-old-standby-important-useful principle from our "regular algebra" bites the dust in this "weird" clock algebra. "If the product of two numbers is 0, then at least one of the numbers is 0." But here $4 \cdot 3 = 0$. The product of two non-zero numbers is equal to zero. In "regular" algebra, we used this principle to solve equations like this one:

$$(x - 5)(x - 6) = 0$$

You can see (by plugging in) that two solutions are $x = 5$ or 6. That is exactly the same result as you would get in "regular" algebra where the number system is the real numbers. But when we are using this 12-hour clock number system, there's a surprise. Plug in $x = 9$. The result is also 0 because we will get $4 \cdot 3$, which equals 12 or 0 on the clock. On this clock, this equation has three solutions, not two, as we would expect.

Mathematicians get all excited when unusual things like this happen. They ask, "What other things about algebra change on this clock?" They wonder what it is about a 12-hour clock that makes this happen. "Does it happen on a 17-hour clock?" (It doesn't.) "Then, what is it about 17 and 12 that causes things to be different?" You get the flavor of how "abstract" algebra can get. Many of us love it!

Boolean algebra (named after George Boole, (1815–1864) is a really weird algebra that uses a two hour clock. The only numbers are 0 and 1. Put 0 at the top and 1 at the bottom. The number facts include these: $0 + 0 = 0$ and $0 + 1 = 1$. But we also get $1 + 1 = 0$ and $1 \cdot 1 = 1$.

It turns out that there are two very useful ways to interpret the 0's and 1's. One is in elementary logic, where 0 means False and 1 means True. The other is in electric circuits where 0 means the switch is open (light switch is Off) and 1 means the switch on closed (light switch is On). Much of the guts of a computer deals with bits (little 0's or 1's) and so Boolean algebra is useful in the design of computer circuits.

Courses Beyond Algebra

Trigonometry

Trigonometry explores the relationship of angles in a triangle with the lengths of its sides.

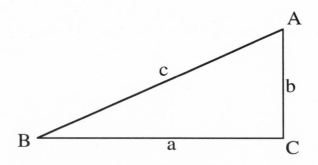

The concept is surprisingly simple. The ratios of all possible pairs of sides depend on the size of angle B, So your Real Trigonometry Book will define functions of angle B that are equal to those ratios. For example, the ratio of side b over side c tells the value of sin(B), pronounced "sine of Bee."

Because there are only six possible ways to make ratios in a triangle, there are only six trig functions. They correspond to the buttons on your calculator: sin, cos, tan, cot, sec, and csc.

You'll like this course, because it'll make more sense than many parts of algebra.

Analytic Geometry   Analytic or coordinate geometry studies how the algebraic equation of a function is related to the geometric shape that its graph has. You were doing analytic geometry when you graphed $y = mx + b$ and discussed how the "m" and the "b" control the positioning of the line.

You will see graceful curves with names like parabolas, hyperbolas, and ellipses in the Cartesian coordinate system. You will also study polar coordinates where you'll use odd looking symbols instead of (x,y). You will graph beautiful heart-shaped curves called cardioids. This area is traditionally considered good preparation for calculus because it requires that you have an integrated understanding of how algebra and graphing work together.

Calculus   In algebra, you found the slope of a *straight* line. Differential calculus studies the slope of a *curved* line.   In the first ten weeks of calculus you learn how to find the "derivative" of a function. A derivative is a new function, based on the original, that tells you the slope of the curve at a point. In algebra, you plot points and connect the dots. In calculus you can plot the point and find the slope at that point. You will often find the derivative of a function, set it equal to zero and then

solve. The solutions you get will indicate the places on the graph of the function where the slope is zero, which is to say, horizontal. These locations are good candidates for finding the maximum or minimum values of the function.

Integral calculus shows how to find the area of a region whose boundaries are described by graphs of functions.

Calculus has a reputation for being difficult. Not true. The ideas and concepts of calculus are easy. Unfortunately, many students have difficulty executing the algebraic steps needed to carry out the details. They learned algebra by memorizing meaningless steps, passing the test and forgetting it all. When they get to the calculus course, they discover that they never really understood what they were doing—or why—in algebra. Many calculus students say they really learned algebra during their first course in calculus. You won't have that problem.

The Problem With Paper

When humans developed paper, they affected the direction their mathematics and science would take for the next few thousand years. Every idea and concept had to be translated into a two-dimensional medium before it could be communicated widely. Paper molded our thought processes and our civilizations.

Reality is larger than that. Objects have height, and weight, and movement. They change with time. They are subject to the tensions of inertia and gravity. They exert their own gravitational and magnetic influences on their surroundings. Some real objects (humans, for example) even think, and feel, and change their minds. All of this must be translated into goose-blood lines on papyrus sheets.

Humans divided this huge chore into several smaller jobs. The poets and historians used words to communicate the thoughts and feelings of human life. Artists created lines of perspective to represent the visual world. Musicians developed a complex system of notes, rests, and crescendo markings to represent sounds.

Philosophers tried to make sense of the universe and man's part in it. Why do things behave the way they do? What forces are at work? And why does man behave the way he does?

Philosophers used words, much as the poets and historians did. They created analogies to explain truths.

A new branch of philosophy began to evolve several thousand years ago. These philosophers described the workings of things by translating them into numbers. By describing something numerically, they could make predictions. If 3 apples plus 2 apples equals 5 apples, then perhaps 3 arrows plus 2 arrows equals 5 arrows. If the stone from a child's sling-shot follows a certain curve, perhaps the boulder in our catapult will follow a similar curve. As if by magic, one could aim a catapult.

This mathematical branch of philosophy became known as science. Rather than describe phenomena in words and analogies, scientists described them using algebra. Science as we know it crystallized about 500 years ago into a style of thinking that boils down to this: Observe something, describe it in terms of symbols and equations, solve the equations, and use the results to predict future behavior. Then experiment to see if you get the results your math predicts.

Human mathematics, science, poetry and music were all shaped by the fact that we make marks on two dimensional surfaces, like cave walls and paper. Imagine how poetry might look on some distant planet where inhabitants write in three dimensions within cotton-candy filled boxes. Poems might rhyme from top to bottom, as well as from side to side. Imagine art on a planet where the medium was in constant motion, like a river running past the artist. And imagine math in a lightless world, where no one could see graphs or equations.

The other forms of communication have all grown beyond paper. Stories now become movies, complete with colors and sound and motion. In the nineteenth century a big musical "hit" meant that people all over the country were buying your sheet music and playing your tune on their piano. Now, we can hear a reproduction of your actual voice singing through our CD player.

Although scientists now use computers to manipulate their equations, algebra itself remains firmly captive to the flat surface that spawned it. Its goal has always been to capture with symbols and numbers the relationships that had previously been expressed in only in words. This process, called the syncopation of algebra, began with Diophantus in 250 AD. and was largely completed by Descartes' time. It has proven to be a powerful and useful development in the growth of mathematics. But it's still just an attempt to express a wildly spinning, colorful reality with lines of goose blood on papyrus.

The language of mathematics is vast and growing. There is no particularly logical place to end our conversation about it except to stop when we grow sleepy, or when we've covered most of the concepts you'll have time to practice in your first year.

And we're sleepy.

The following comments were sent by students of **Dr. James H. Snider's** 8th grade Algebra I class at East Middle School in Nashville, Tennessee:

"I learned that algebra is a tool kit for solving mysteries using whatever clues we can find to solve the problem. I really thought this book was pretty cool." **William Brewner**

"Well, I learned some from this book. Algebra sounds a lot like chess because of the moves and strategies to be learned." **James Maes**

"I can do my homework with less problems because now I have more strategies to use and I can do them quicker and with more ease because I understand them a lot better than I did before I read *Algebra Unplugged*." **Dan Emery**

"This book was not only an interesting way of getting more familiar with math, but it also sent out a strong message: telling kids that algebra can be fun!" **Crystal Jones**

"The book is helpful in the way that algebra is compared to real life, but confusing the the sense that it skips from one topic to another. I didn't realize there was really a strategy to algebra like there is in chess, a sort of divide and conquer. In equations you first get the variable by itself (divide) and then work the other side (conquer)." **Marcus Horton**

"If I didn't read this book, I would have failed this class because all Marcus and I used to do in the class was clown around and talk about how good Tabitha looks. So I would like to thank you for giving me this book to read." **Tyrone Williams**

"*Algebra Unplugged* is geared for the majority who are not good at math. Told with humor and understanding, it is an enjoyable book to read."
Joan Harris, *The Institute for Science and Society*

"I very much enjoyed the book. A math book with lawyer jokes and stories of flogging tenors. It doesn't get any better than that."
Thomas D. Seidenberg, *Phillips Exeter Academy*

"The volume's easy pace and the use of a game as a metaphor probably will appeal to the casual learner. The book's gentle, conversational, gamelike approach may be sufficient to reach the 'unreachable.'"
Science Books and Films,
American Association for the Advancement of Science

"This book discusses algebra in a non-threatening, fun way."
ERIC Clearinghouse for Science, Mathematics, and Environmental Education

"The book contains no exercises. Instead, it simply explains the concepts, vocabulary and strategies of algebra in understandable terms."
Zentralblatt für Didaktik der Mathematik

"Sometimes, despite endless explanations by teachers and dozens of homework assignments, students don't always grasp algebra. Some ask for help, others turn to books, hoping that one will explain things in language they can understand. This may be the book they are looking for. Explanations are short, humorous, and non technical. The authors convinced this reviewer that there is value in sneaking up on a potentially intimidating subject in this way, although I was not so sure at the beginning."
Appraisal– Science Books for Young Adults

"Ever read an algebra book for fun? Ever thought you would want to? *Algebra Unplugged* is just that sort of book–an innovative approach that does a great job taking the mystery and fear out of algebra. It's not a text book. You don't have to ever lift a pencil while reading it. Fascinating explanations of all the players in a first year algebra course. A must for anyone who is going to take algebra, dreaded it while taking it, or wants to brush up on it."

Theoni Pappas, author of *The Joy of Mathematics*

(More comments on preceding page)

Ordering Information
Available at book stores everywhere.

Distributed to the book trade by Ingram Book Company, Baker and Taylor, and Quality Books.

Individual copies of *Algebra Unplugged* may be purchased directly from the publisher for $17 including shipping.

Individual copies of *There Are No Electrons: Electronics for Earthlings* by Kenn Amdahl may be purchased for $15 including shipping.

We'd love to hear your comments, and may include some in the next printing.

Clearwater Publishing Co
P.O. Box 778
Broomfield, CO 80038-0778

Contact the authors by e-mail at:

Kenn Amdahl: Wordguise@aol.com.
Jim Loats: loatsj@mscd.edu